Rômulo D. C. Santos, Ph.D.

Numerical Modelling and Simulation of Fractals in C++

Rômulo D. C. Santos, Ph.D.

Numerical Modelling and Simulation of Fractals in C++

Geometric Modeling of Fractals with Object-Oriented Programming Language

ScienciaScripts

Subproject Summary

INTRODUCTION: In Euclidean geometry, the simplest and best known figures are studied, such as: straight lines, squares, circles, cones, pyramids, among others, and it is common to calculate their measures of length, area and volume. In this context the idea of dimension is already in our minds, so many phenomena and forms found in nature can not be explained in the ways of conventional mathematics, requiring a special theory to explain them and characterize the so-called fractal geometry. According to (TRICOT, 1955), fractals come from the Latin *fractus* which means "broken", they are geometric shapes with some special characteristics that define them and distinguish them from other shapes, such as self-similarity at different levels of scale, a characteristic that does not exist in all fractals. Currently the fractal geometry, and in special the fractal dimension, has been used in several areas of knowledge, such as the study of chaotic systems, cloud formation patterns, object characterization, analysis and recognition of patterns in images, texture analysis and measurement of curve lengths. According to (DEVANEY, 2004), new ideas as Riemannian modeling for construction of "new structures" as fluid particles, in which these in turn, support us to create and model fluids such as water, smoke and gases from some constructions and a good physical-mathematical basis. **MATERIAL AND METHODS:** Firstly, all the theoretical reference about fractals knowledge was gathered from books at the library of the Universidade Federal do Acre - UFAC (Federal University of Acre) and researches on the internet. Then, the need to use and study a programming language to build fractals was verified. To build the algorithms it was used 3D *free software* (*Blender*® *and Studio Max*), focused on applying the concepts of computer graphics. Some seminars were held in the Mathematics Degree Course, approaching the research content, and for the implementation and use of the programs, the Informatics Laboratory of the Mathematics Degree Course at UFAC was used. **RESULTS:** As result it was verified that it is possible to apply the mathematical concepts to show the abstract behavior of certain functions like: Complex Functions (vector calculus in the space R^n , Frenet curves, etc.), as well as in Geography the demographic growth models. With Computer Graphics, it was possible to visualize, create methods and develop 3D objects, allowing the introduction of a new field of study, that is, the modeling of Riemannian Geometry and the behavior of these "new structures" of fractals and their consequences, such as the *modeling of fluids,* a very significant result, in which it was possible to start from an idea like fractals, create new structures and new models. **CONCLUSION:** With this research we conclude that fractal geometry helps to bring together natural sciences and computer graphics, along with mathematical concepts and programming languages, allowing the topological representation of fractals through an object oriented programming structure in multiplatform languages. Thus it is believed that this work will contribute to future studies of modeling in various sciences using computer graphics.

KEY WORDS: Fractals, Computer Graphics and Mathematical Modelling.

Introduction

With this project, a new construction of knowledge is proposed - Computer Graphics - as the key for the modelling of a new geometry; with a high level language, but at the same time didactic and of free access. The construction of new fields and new methods for the scientific knowledge and its contribution for several segments.

Until the early 19th century, the geometry presented in Euclid's *Elements* and developed by later mathematicians was believed   to be the only possible geometry. This geometry was also believed to be *true*, in the sense that it corresponded to reality.

To the amazement of mathematicians, as a result of these attempts to prove the postulate of parallels, the first non-Euclidean geometries were created in the 19th century, starting with the work of Lobachevsky, Bolyai and Gauss, followed by important developments such as the creation of non-Euclidean differential geometry and the study of spaces with more than three dimensions. The development of these geometries gradually led to the realization that Euclidean geometry was only *one* of several possible (not contradictory) geometries, and that all of them were equally valid, none of them being *truer than* the other.

With respect to non-Euclidean geometry, we started from Bonola's book (1955) which presents general aspects of the development of this geometry from the sixteenth century until the theories of Lobachevsky and Bolyai. This text also provided the first bibliographic references, both primary and secondary, directly related to the subject.

Today it is possible to distinguish Euclidean and non-Euclidean geometry because both are known and have their clear differences when compared. But it is clear that when only Euclidean geometry was known, there was no reason to characterise it as "Euclidean geometry", because it was the only one. Thus, the name "non-Euclidean geometry" only became popular when the possibility of constructing geometries different from the old and which violate Euclid's postulate was realised. The awareness that the new geometry would be a non-Euclidean geometry only happened some time after the first proposals. Only after this awareness does it make sense to speak of Euclidean and non-Euclidean geometry.

Objectives
General

Perform the topological representation of fractals through object-oriented programming framework in a high-level and cross-platform language.

Specific

- Theoretical knowledge in pure mathematics: derivation and numerical integration;

- Knowledge of geometric optics;

- Knowledge of Euclidean geometry and Riemannian geometry;

- Properties of graphic objects in dimension one, two or three;

- Geometric Modeling of Fractals with algorithms in C, C++, Pascal, JavaScript or Python languages;

- Computer Graphics Object Modeling in 3D environment with free software.

Material and Methods

At first, after bibliographical survey, to give all the theoretical support. The scholarship holder will be able to start the development of the research and every three months present the progress of the project to students of Mathematics and Information Systems, since the subject is common to both courses.

Use of free 3D software as the main support tool, for the modelling of Fractals and their respective Non-Euclidean Graphical Modeling of Surfaces with object-oriented programming language.

Results and Discussion

The results to be expected are the application of the tools of Pure Mathematics in which it uses abstract behaviour in the structure of certain functions. In Computer Graphics, it will be possible to visualize, create methods and develop objects in a 3D environment, allowing the introduction of a new field of study, that is, the modeling of a non-Euclidean geometry and the chaotic behavior of fractals. We can observe some results obtained, such as the modeling of fractals resulting in the creation of particles, which in turn, can generate various forms of "objects" in nature, such as terrain, clouds, water, among other fluids. That's right! We will present significant results with the use of Physics - Mathematics, as "key" for the construction of scientific knowledge and respective results in the computational scope related to the 3D environment.

Conclusion

The study of Geometric Modeling of Fractals with Object Oriented Language reveals itself so far, only as the "tip of an iceberg". Characterized of a great importance for the knowledge areas. The geometric modeling of fractals has relevant methods, when it comes to modeling "objects" of nature, such as fluids, terrain, clouds, etc.. The geometry of chaos, although a subject not much explored in our environment, but only addressed in a superficial way, becomes to have an importance of great interest in solving problems related to the study of climate and weather, the study of the behavior of fluids and their dynamics, among other problems where we can not find exact and precise solutions. After this study we start to look at the non-Euclidean geometry as a geometry that brings in its structure, answers to various problems not yet solved by a more accurate and rectilinear geometry.

Some applications are quite relevant in the study of fractals, for example, in Geography the description and characterization of seismic faults and, consequently, earthquakes are obtained through the study of their fractal structure. Besides earthquakes, other geological phenomena can be studied such as, for example, the dynamics of volcanoes. In economics, the analysis of global behaviour helps to understand local behaviour and vice versa. The study of these analyses makes it possible to create medium and long-term commercial strategies.

Another quite important result is the multidisciplinary nature of fluid animation. It is another aspect to be highlighted. The development of this area depends on the interaction between the areas of computation and fluid mechanics. This fact brings a singular dimension to this area of computational animation. On one hand, animation specialists seek in the fluid modeling techniques, tools to meet their needs in computer graphics. The final rendering of the scenes depends on the development of resources in scientific visualization which, in turn, may be useful to engineers in the analysis of the results of their simulations. Therefore we can glimpse a very fertile field of work with ample possibilities for research.

Bibliographic References

BOLYAI, John, *The Science of Absolute Space*, English translation by George Bruce Halsted, in BONOLA, Roberto, *Non-Euclidean Geometry*, English translation by H. S. Carslaw; New York: Dover Publications Inc, 1955.

BONOLA, Roberto, *Non-Euclidean Geometry,* English translation by H. S. Carslaw; New York: Dover Publications Inc, 1955.

DAVID, Ebert. **Texture and Modeling**: *A procedural Approach*. Academic Press, 1994.

______. Procedural Volumetric Modeling and Texturing, **SIGGRAPH 97**, course 14, notes, chapter 6, August 1997.

DEVANEY, R. L., Differential Equations, Dynamical Systems, and an Introduction to Chaos, second edition, Elsevier Academic Press Inc.

FOSTER, Nick; METAXAS, Dimitri . Realistic Animation of Liquids. **SIGGRAPH 99**, *course notes*. August 1999.

GOODMAN, Danny. Java **Script: The Bible. Translation: Daniel Vieira**. 3ª triagem. Rio de Janeiro: Elsevier, 2004.

GUIDORIZZZI, Hamilton Luiz. **Um Curso de Cálculo**. 4. ed. V. 1, 2, 3 e 4. Rio de Janeiro: LTC, 1997.

LOBACHEVSKY, Nicholas, *The Theory of Parallels*, English translation by George Bruce Halsted, in BONOLA, Roberto, *Non-Euclidean Geometry*, English translation by H. S. Carslaw; New York: Dover Publications Inc, 1955.

Mandelbrot, B. The Fractal Geometry of Nature.

W.H. Freeman, New York, 1983.

MUSGRAVE, F. Kenton; KOLB, Craig E. The synthesis and rendering of eroded fractal terrains. **Computer Graphics (SIGGRAPH '89Proceedings)**, v. 23, p. 41-50, July 1989.

NEYRET, F., Qualitative Simulation of Conventional Cloud Formation and Evolution. *Eight International Workshop on Computer Animation and Simulation,Eurographics*, September 1997.

Blender 3D free software. Available at: < www.bender.org > Accessed on: 20 nov. 2007.

TRICOT, C. Curves and Fractal Dimension. Springer-Verlag, New York, 1995.

WYVILL, G. ; MCPHEETERS, C.; WYVILL, B. . **Data Strucuture for Soft Objects, *The Visual Computer***, v. 2, p 227-234, January 1986.

Evaluation of the Scholar

According to the work plan initially proposed, it can be seen that the scholarship holder was able to develop all the proposed steps to her satisfaction, and it was also possible to program three more methods that were not part of this research. With this, it already shows all its commitment, responsibility and ability in the proposed area.

Advisor's Evaluation

SUMMARY

1. INTRODUCTION

At first, we must make a small trip back in time; to be more precise, through the history of mathematics. We will try to build a single logical-mathematical reasoning, reasoning capable of building and making new connections between the field of "mathematical modelling", "Euclidean geometry", "non-Euclidean geometry" and "computer graphics".

The latter portrays the "virtual world" as the "best" environment for the presentation of new "structures" of spaces and the visualization and construction of "new geometries". What we seek here is to present "new ways" to outline Euclidean and Non - Euclidean Geometry, and use these principles to develop a language; object oriented language; where we can build new mathematical structures capable of revealing a "new world". A world that is unknown to many. Under a new aspect, under a new environment of production and visualization of results, we can apply and take these concepts about mathematical modeling in computer graphics to classrooms and computer labs to students of medium and higher levels; promoting a "strong and applied" learning. And together with that, to the professional improvement. Computer graphics as a bridge between modern mathematics and the virtual world.

We want to propose here, a new construction of knowledge - Computer Graphics - as a key to modeling a new geometry; with a didactic language and free access to all. The construction of new fields and new methods for scientific knowledge and its contribution to society.

Until the early 19th century, the geometry presented in Euclid's *Elements* and developed by later mathematicians was believed to be the only possible geometry. This geometry was also believed to be *true*, in the sense that it corresponded to reality.

Although the validity of this geometry was not in doubt, there was some dissatisfaction with its foundations, because of the non-intuitive character of the postulate of parallels. Since antiquity, several authors (such as Proclos) have tried to replace the postulate of parallels by a more intuitive one, or to demonstrate it from the other postulates, in order to make the geometry better founded.

To the amazement of mathematicians, as a result of these attempts to prove the postulate of parallels, the first non-Euclidean geometries were created in the 19th century, starting with the work of Lobachevsky, Bolyai and Gauss, followed by important developments such as the creation of non-Euclidean differential geometry and the study of spaces with more than three dimensions. The development of these geometries gradually led to the realization that Euclidean geometry was only *one* of several possible (not contradictory) geometries, and that all of them were equally valid, none of them being *truer than* the other.

The emergence of non-Euclidean geometries had strong repercussions in physics, for two reasons. Firstly, because until the early nineteenth century many authors believed that Newtonian mechanics was a true *a priori* theory - like geometry - and that its basic principles could be proved from intuitive postulates or axioms. This conception fell apart, together with the analogous view regarding geometry. The simplest alternative, for physicists, was that mechanics was empirical knowledge, but there seemed to be some important difference between mechanics (which was described, until the end of the 18th century, as an area of mathematics) and other disciplines such as the study of electricity or heat (which were accepted as empirical studies). Gradually, the suspicion began to arise that there might be several different mechanics.

At the same time, the existence of different geometries had an immediate consequence for physics, because many physical theories use geometry (together with physical assumptions) in their deductions. Even if the basic principles of physics (including those of mechanics) were maintained, the existence of different geometries opened up the possibility of alternative physical theories leading to new consequences by the use of these alternative geometries.

Within mechanics a particular principle was the subject of controversy during the 19th century and later ended up being directly influenced, and also influencing, the different geometries that emerged. The principle of minimum action had metaphysical origins, with Leibniz and Maupertuis, but with the introduction of Euler's variational calculus, it was formulated mathematically and made it possible to determine the equations of motion. However, the expression found by Euler was not directly associated with action. Lagrange, in turn, defined what he called the *principle of minimum action* and determined the equations of motion from the extreme condition, where the variation was null. The way Lagrange deduced the principle of least action allowed us to find the equations of motion, very similar to what we have today, but did not directly provide the equation of the trajectory.

The relation between the principle of least action and the equation of the trajectory appeared in the work of Poisson. He finds a condition in which the principle of least action, in the form given by Lagrange, could be interpreted as the equation of the geodesic. However, the form that Poisson gave to the principle did not consider the position-dependent forces, which made his result restricted.

Another way to obtain the equations of motion comes from the relationship that Hamilton makes between optics and mechanics. But Hamilton's theory also did not allow us to directly find the equations of the trajectory and did not explicitly mention a relationship with the principle of minimum action. The final relation, which geometrized mechanics, only appeared with Jacobi, who rewrote the principle of least action by eliminating the time variable, and then directly finding information about the trajectory of a body subject to the action of position-dependent forces. Jacobi's geometrization of mechanics allows a new interpretation of Newtonian mechanics, because in this new formulation it is possible to find the equation of the trajectory without considering concepts like velocity, acceleration, etc.

From the middle of the 19th century, differential geometry became the language adopted in the description of the theories of heat and elasticity, which Lamé does using differential parameters. In the same period, solutions to problems with n degrees of freedom in mechanics emerged, which led to *n-dimensional* differential geometry.

Jacobi's interpretation of the principle of minimum action depended exclusively on a quadratic form, which led to the relationship between this principle and the determination of the equation of the geodesic. After the publication of Riemann's work in 1867, the metric became the main characteristic of space, determining its properties. This made the principle of minimum action have other interpretations in an *n-dimensional space.*

This relation between the principle of minimum action and non-Euclidean geometry, that is, the determination of the geodesic in any *n-dimensional* space, appears clearly in the works of Darboux and Lipschitz, who dealt with *n-dimensional* mechanics. But the mathematical formalism which made it possible to interpret this relationship in the way that was used in the theory of relativity was the absolute differential calculus, which appeared at the end of the 19th century.

Although we have restricted ourselves to the 19th century in the main part of the work, it was often necessary to return to earlier authors so that we could understand the sequence of facts. Especially regarding the principle of minimum action, the work would be incomplete if we did not return to the seventeenth and eighteenth centuries to understand the definition given by Leibniz and Euler, because we did not find secondary literature that could clarify the differences with respect to the principle as we know it today, or even the difference between the principle of Lagrange and Hamilton.

The absence of works that pointed out the differences in this principle and its direct relationship with the metric of space led us to questions that we tried to answer throughout the work. How did the geometrization of mechanics happen and how was this fact related to a principle of metaphysical origin, like the principle of minimum action? How was the interpretation of this principle made in the *n-dimensional* case? What were the consequences, before the scientific community, of the emergence of a geometry different from Euclidean geometry? When this geometry began to be applied in the physical sciences? These are some questions that led the work.

As part of the research methodology adopted in a work on the history of science, the research had as a starting point a secondary text that showed a general idea about the main subject of the thesis. The relationship between geodesics and non-Euclidean geometry was already well known from the theory of relativity, but there were no works that delved into how this relationship began, what the initial hypotheses were, etc.

With respect to non-Euclidean geometry, we started from Bonola's book (1955) which presents general aspects of the development of this geometry from the sixteenth century until the theories of Lobachevsky and Bolyai. This text also provided the first bibliographic references, both primary and secondary, directly related to the subject.

2. EUCLIDEAN AND NON-EUCLIDEAN GEOMETRY

Today it is possible to distinguish Euclidean and non-Euclidean geometry because both are known and have their clear differences when compared. But it is clear that when only Euclidean geometry was known, there was no reason to characterise it as "Euclidean geometry", because it was the only one. Thus, the name "non-Euclidean geometry" only became popular when the possibility of constructing geometries different from the old and which violate Euclid's postulate was realised. The awareness that the new geometry would be a non-Euclidean geometry only happened some time after the first proposals. Only after this awareness does it make sense to speak of Euclidean and non-Euclidean geometry.

Non-Euclidean geometry has its origin in Euclid's postulates, that is, the foundations of Euclidean geometry. Euclid's *Elements* are the basis for the development of an axiomatic geometry, founded on definitions and postulates that have served, and serve, as a reference for the study of plane geometry.

However, one of Euclid's postulates, the fifth postulate, also known as the *postulate of parallels,* aroused doubts which led to a geometry which can be constructed like Euclid's, without presenting contradictions, but which leads to different theorems. For centuries, or more precisely since the first known attempt at a proof by Geminus in the 1st century BC, several scholars have tried to demonstrate Euclid's fifth postulate, without results, but the various hypotheses ended up serving as a starting point for the geometries of Lobachevsky and Bolyai, established in the 19th century.

Between these 20 centuries, geometry went through periods of little and great interest. With the translation of Euclid's book into different languages, the fifth postulate was the object of study among the Arabs and then among Europeans, who made new attempts at demonstration.

Saccheri's attempt at demonstration by absurdity in 1733 may be considered the first to show the possibility of constructing a new coherent geometry different from that postulated by Euclid, although Saccheri himself did not see it.

The introduction of a new geometry only really came about with the works of Lobachevsky (1829) and Bolyai (1832). Still, these discoveries were not enough for a generation to convince that a geometry other than Euclid's was possible.

Only in the 1960s did the mathematical community begin to accept the new geometry.

There is another development in geometry that occurs parallel to the emergence of these works, which studies curved surfaces, and which is not trying to change Euclidean geometry. Within this line of research, Gauss's work on differential geometry in 1827 shows the possibility of analysing surfaces using intrinsic properties. The connection between Gauss's work and the existence of a new geometry could only be perceived after Riemann's work in 1854 (published only in 1867), but the relationship between this new geometry and the fifth postulate took time to happen, because differential geometry presented another way to study space.

The first to employ the term "non-Euclidean geometry" was Taurinos in 1825, with the creation of logarithmic-spherical geometry, but it did not get much attention.

The works of Lobachevsky and Bolyai are contemporary to Gauss's work on curved surfaces and are complementary. While the former developed a new geometry axiomatically, Gauss exploited Euclidean geometry and spherical trigonometry to describe surfaces with a formalism that depended only on their intrinsic properties. With Riemann's work, metrics and curvature of space became the main way of describing *n-dimensional* space, and Euclidean geometry was the limit for the three-dimensional case.

Considering these different approaches, one can have two types of non-Euclidean geometries: those limited to three-dimensional space, but with non-Euclidean properties, as is the case of Lobachevsky and Bolyai; and those dealing with a space with dimension greater than three.

In this chapter, we will address from the beginning of the questions about the postulate of the parallels, until the geometry developed by Lobachevsky and Bolyai, limiting ourselves to the first type of non-Euclidean geometry. We will not discuss in detail the mathematical proofs of each demonstration, because it is not part of the goal of this paper and they can be found in the work of Bonola and Gray (BONOLA, 1955; GRAY, 1989).

2.1 - EUCLID'S FIFTH POSTULATE

Euclid's *Elements* consists of 13 books, in the version we know, which contain the basics of geometry and arithmetic. Both the history of the books and that of Euclid himself (~325 BC -~265 BC) is full of contradictions and speculations that have arisen from the many translations they have had over the centuries2. In this paper we will devote ourselves only to book I, with the comments made by Proclus (410-485).

In this version, Book I has 23 definitions, 5 postulates, 5 common notions (axioms) and 48 propositions (theorems). The definitions and axioms are undemonstrated, and the common notions are non-geometrical general principles. Like axioms, postulates are undemonstrable and must be admitted. The propositions, or theorems, are proved from the previous definitions, postulates, common notions and theorems by the deductive method.

At the beginning of book 1, Euclid provides some definitions, of which we will highlight three of greatest interest for the work, which are assumed to be true and which define concepts that will be part of the postulates. He defines straight and parallel lines as follows (EUCLIDES, vol. 1, 1956, pp. 153-154):

2. A line is a length without width;

4. A straight line is a line that has the points on it placed regularly;

23. Parallel straight lines are straight lines which, being in the same plane and extended indefinitely in both directions, do not meet in any direction.

The postulates determine the mode and possibility of construction, that is, what is possible to construct from the initial definitions. The postulates in this version of book I are (EUCLIDES, vol. 1, 1956, p. 155):

You can] draw a line between any two points.

It is possible to extend a finite line continuously into a straight line.

It is possible to draw a circle with any centre and any radius.

4. All right angles are equal to each other.

5. If a straight line that cuts two other straight lines forms interior angles on the same side whose sum is less than two right angles, the two straight lines, if extended indefinitely, will meet on the side where the sum of the angles is less than two right angles.

That is, let *r* and *s be* coplanar lines cut by a third line *t*. If the sum of the interior angles a and b on the same side is less than 180°, then the two lines *r* and *s* will meet when extended to that side (see Figure 2.1).

In some presentations of Euclid's work, the way of placing the postulates and axioms is different, and what is usually called the fifth postulate appears as Axiom XI.

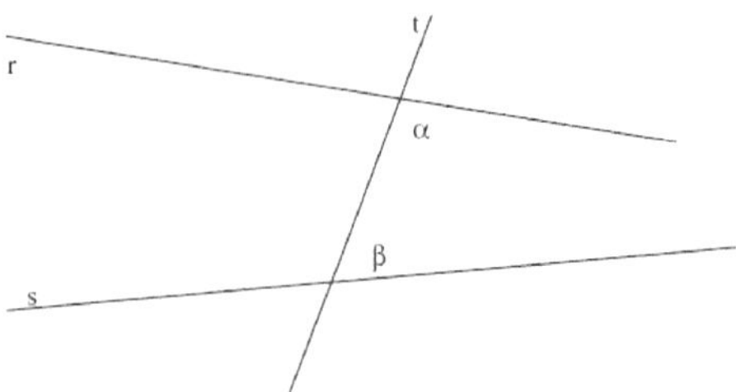

Figure 1: According to the fifth postulate, the lines *r* and *s* meet on the side where the interior angles α ☐E β☐somam less than 180°.

From the definitions and the first four postulates, Euclid demonstrates the first 28 propositions and begins to use the fifth postulate from proposition 29 onwards. There are several important propositions which can only be demonstrated using the fifth postulate, such as (EUCLIDES, vol. 1, 1956, pp. 311-322):

Proposition 29: A straight line cutting two parallel straight lines makes the alternate angles equal to each other, the exterior angle equal to the opposite interior angle, and the interior angles of the same side equal to two right angles.

Proposition 30: Straight lines parallel to the same straight line are parallel to each other.

Proposition 31: Through a given point passes a straight line parallel to a given straight line.

Proposition 32: In any triangle, if one of the sides is extended, the external angle is equal to the two internal and opposite angles, and the three internal angles of the triangle are equal to two right angles.

Proposition 33: Straight lines joining equal and parallel (at their ends) straight lines in the same directions (respectively) are also equal and parallel.

The fifth postulate is also used to establish the properties of similar figures (BONOLA, 1955, p. 2).

Although people did not doubt that the geometry presented in the *Elements* was correct, they considered that the fifth postulate was not intuitive and was much more complicated than the others. Attempts then arose to demonstrate the fifth postulate, so that it would become a proposition, or should be replaced by another, simpler postulate.

In order to try to demonstrate the fifth postulate, other definitions were given for "parallel lines" during antiquity. The definition given by Posidonius of Rhodes (135-51 BC), as quoted by Proclus in his comments, is based on the distance between the lines and the perpendiculars that can be drawn:

> Parallel lines are those which, being in a plane, neither converge nor diverge, but have all perpendiculars which are drawn from a point of one of the lines to the other, equal. (EUCLIDES, vol. 1, 1956, p. 190)

But it is not possible to assert that if two straight lines, supposed to be parallel, are cut by a third line, making a right angle with one of the parallel lines, it necessarily makes a right angle with the other parallel line. This statement is only possible if we consider the fifth postulate. Thus, this definition of parallels assumed the fifth postulate to be true, which generated a contradiction. Gemini (1st century B.C.) used Posidonius' definition of parallel to try to demonstrate the fifth postulate, but ended up using the proposition, which invalidated his demonstration, as did Claudius Ptolemy (2nd century A.D.) (EUCLIDES, vol 1, p. 297).

The fifth postulate has been the subject of study for many centuries, in different periods of history. From the attempt by Ptolemy to John Wallis (1616-1703), Euclid's *Elements* were known to various peoples and the fifth postulate was the subject of attempts at demonstration using the theory of equidistance, similarity of figures, etc., which in general only determined another way of stating it (BONOLA, 1955, pp. 1-21).

2.2 - FIRST ATTEMPTS AND ARABIC WORKS

During the Middle Ages, in the 9th and 10th centuries, the postulate was studied more by Arabs than in Europe. With the translations of Euclid's books among the Arabs, some works tried to demonstrate the fifth postulate. Among the Arabs, the works of Al-Gauhari, Thabit ibn Qurra, Ibn al-Haytham, Omar Khayyam and Nasir Eddin al-Tusi6 stand out.

In his second treatise on Euclid's geometry, Thabit ibn Qurra (836-910) initially established that a line segment could move without having its length changed, because he considered that the absence of deformation of geometric figures when there is movement was not obvious (GRAY, 1989, p. 43).

The introduction of motion for the measurement of magnitude was an idea that had been challenged by Aristotle7 and Ibn Qurra argued that Euclid himself had used it when demonstrating proposition 4 of book 1 (ROSENFELD, 1988, p. 52). In the demonstration of proposition 4, Euclid considers that one of the triangles can be moved and placed over the second to make his analysis. Ibn Qurra introduces a figure that will be widely used in attempts to demonstrate the fifth postulate: a quadrilateral with two equal and opposite sides making the same angle with the base, as in Figure 2.2, where sides DA and CB are equal, and angles $B\hat{A}D$ and $A\hat{B}C$ are equal.

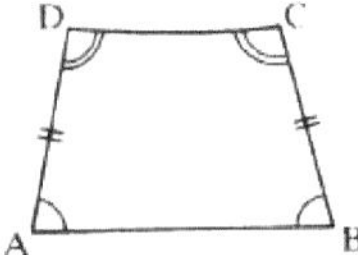

Figure 2: Ibn Qurra considered a quadrilateral with equal angles at the base and equal sides AD and BC

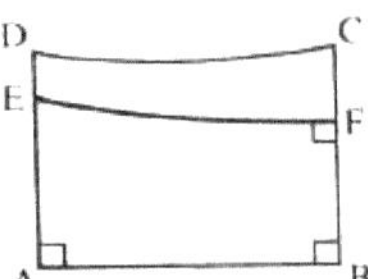

Figure 2.1: The angles of the base are right angles EF equals AB

According to this hypothesis, in a quadrilateral in which the base angles are right angles, as in Figure 2.3, and considering *EF* equal to *AB,* since this line would be the offset of the base and therefore maintain its length, then the fourth angle, *AÊF*, would also be right angles. Using this hypothesis, Ibn Qurra tried to prove the fifth postulate using figure 2.4.

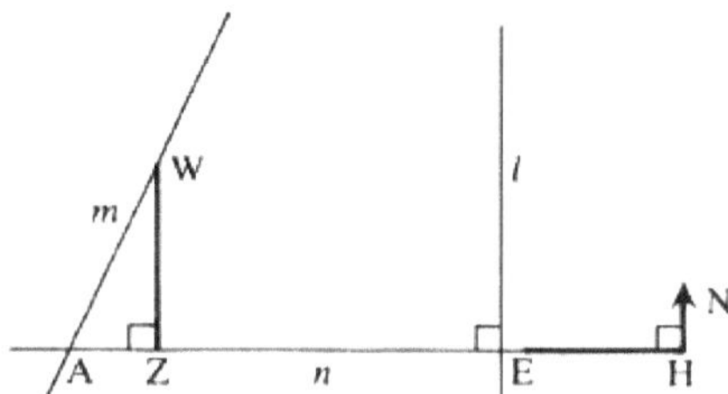

Figure 2.2 - Ibn Qurra's construction to try to demonstrate the 5th postulate

In the figure, lines *l* and *m* cut a third line *n*, where *l* and *n* are perpendicular (hypothesis). Then he took a point *W* on *m* and drew the perpendicular *WZ* to *n*. If *AZ* is smaller than *AE*, then some multiple of it, *AH* exceeds *AE,* and can form the right triangle *AHN*, where *N* is on *m*. According to the quadrilateral hypothesis, NH does not find *l*, for if it did, it would not form the quadrilateral he initially introduced (figure 2.2). Then *l* must meet *m* and the fifth postulate is proved, that is, the lines *l* and *m* meet on the side where the sum of the angles (angles Â and Ê) add up to less than two right angles.

However, in his demonstration Ibn Qurra had to assume one of the propositions that used the fifth postulate in the demonstration10, which invalidated his demonstration (GRAY,1989, p. 44).

Nasir Eddin al-Tusi (1201-1274) also used the quadrilateral with at least two right angles and reasoning similar to Ibn Qurra's to make his demonstration of the fifth postulate (ROSENFELD, 1988, pp. 74-85). However, he failed to suppose that lines that seemed to converge on one side would effectively meet. His demonstration caused the greatest interest in Europe when published in Rome in 1594 (GRAY, 1989, p. 51) and is the only one that appears in most books on the postulate, such as Bonola (1955).

Other mathematicians made different hypotheses in their attempts at demonstration, but without success, and criticism of the fifth postulate appeared more frequently in translations of the book during the 16th century (ROSENFELD, 1988, pp. 92-107).

2.3 - TAURUS, AN UNKNOWN

Carl (or Karl) Friedrich Gauss (1777-1855) was a brilliant mathematician, who made contributions in the areas of number theory, function theory, probability, geometry, etc.

By some historians, he is considered one of the founders of non-Euclidean geometry, because from several of his letters it can be inferred that he was conducting research in this area. However, although he wrote in a letter12 to Wolfgang Bolyai that he had *been "thinking about the subject for 40 years",* Gauss did not publish any work in which he discussed the possibility of the existence of a geometry other than Euclidean geometry (GRAY, 1989, pp. 86-87).

During the 18th century, geometry was based on two different paths: spherical trigonometry, which was used in astronomical studies; and a geometry, in the plane, which was based on Euclid's arguments. It was based on spherical trigonometry that Lambert proposed that the acute angle hypothesis could be associated with a sphere of imaginary radius. The sine and cosine functions were already known in the form given by Euler13 , and with this Lambert associated the properties of a sphere of imaginary radius with hyperbolic sine and cosine.

In the same period that Gauss was researching the parallel postulate, but independently of it, Ferdinand Karl Schweikart (1780-1859) was publishing work on the same subject using spherical trigonometry. In 1807 he published *Die Theorie der Parallellinien nebst dem Vorschlage ihrer Verbannung aus der Geometrie,* in which he developed the fifth postulate based on the idea of the parallelogram, without being able to prove it (ROSENFELD, 1988, p. 218). But in a letter written to Gauss in 1818, he described a geometry that would not be based on Euclid's hypotheses, which he called *astral geometry.*

Astral geometry is constructed starting from the property that the sum of the three angles of a triangle is not equal to two right angles, that is, it can be associated with the acute angle hypothesis:

This being so we can, rigorously prove that:

a) the sum of the three angles of a triangle is less than two right angles;

b) the sum becomes smaller the larger the area of the triangle;

c) that the height of an isosceles triangle increases continuously, as do the sides, but it can never become greater than a certain length, which I call the *Constant* (SCHWEIKART, *apud* BONOLA, 1955, p. 76).

Euclidean geometry would hold in the case where this constant is infinite. Schweikart is clearly influenced by Saccheri and Lambert, since astral geometry possesses the characteristics of the acute angle hypothesis of the former and the relationship between area and angle value of the latter (BONOLA, 1955, p. 77). This also justifies his starting directly from the acute angle hypothesis, without considering the obtuse angle hypothesis, since Saccheri and Lambert had already rejected this hypothesis.

In a Schweikart triangle (Figure 2.3), in which the three pairs of sides are asymptotic to each other, the area is given by $\dfrac{\pi C^2}{\left\{\log 1+\sqrt{2}\right\}^2}$ where C is the constant of astral geometry.

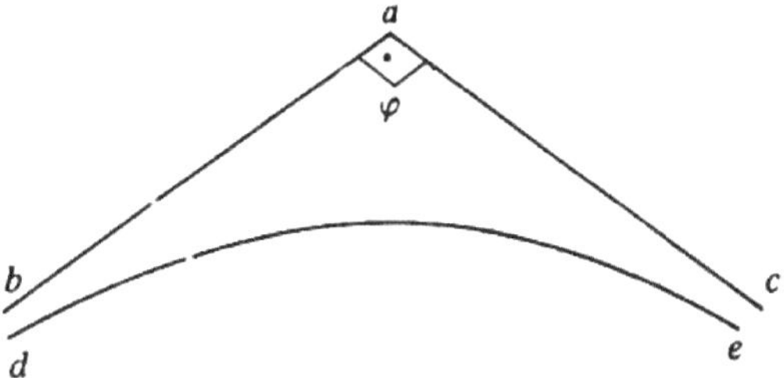

Figure 2.3: Schweikart's triangle, where *ab* and *ac* are asymptotics next to.

Despite receiving a positive response about his theory from Gauss, Schweikart did not continue his work on the subject, convincing his nephew, Franz Adolf Taurinus (1794-1874), who was a mathematician, to make investigations into astral geometry.

Taking a different position from his uncle, Taurinus published two works in which he developed astral geometry14: *Theorie der Parallelinien* (1825) and *Geometriae Prima Elementa* (1826), always having the fifth postulate as an absolute truth. In the first of these, Taurinus rejected the obtuse angle hypothesis, influenced by the works of Saccheri and Lambert, and rejected the acute angle hypothesis, since the validity of the latter would lead to the existence of infinite parameters (Taurinus' parameter corresponds to Schweikart's Constant). The existence of infinite Taurinus parameters would lead to the non-existence of an absolute measure of length, which contradicted his conception of space (BONOLA, 1955, p. 78).

However, in his second work, Taurinus no longer assumes the fifth postulate to be true and develops the acute angle hypothesis, concluding that it would be equivalent to a spherical trigonometry in which the radius is an imaginary number. He named his theory *logarithmic-spherical geometry* and determined its properties by replacing the radius in spherical trigonometry (*k*) by an imaginary number (*ik*). Considering the fundamental formula of spherical trigonometry, depicted by figure 2.4, given by:

$$cos\,\frac{\alpha}{k} = cos\,\frac{\beta}{k}cos\,\frac{\gamma}{k} + sen\,\frac{\beta}{k}\,sen\,\frac{\gamma}{k}cosA$$

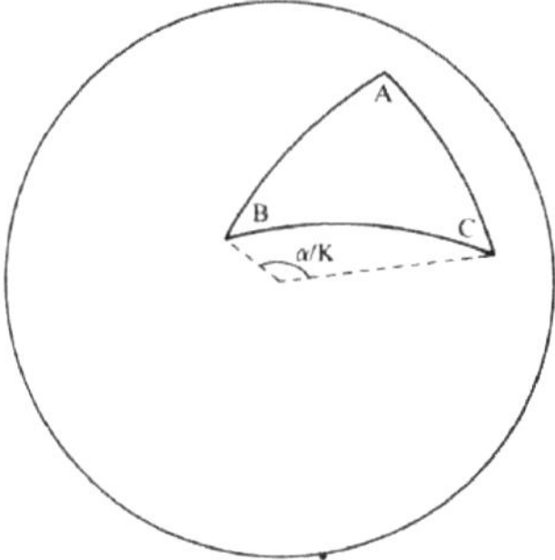

Figure 2.4: Representation of the spherical triangle on a sphere of radius k

Where b is the angle at B, g is the angle at C and a is the solid angle. By replacing the radius *k* with an imaginary radius *ik*, we can rewrite it using hyperbolic functions, such as:

$$cos\frac{\alpha}{k} = cosh\frac{\beta}{k}cosh\frac{\gamma}{k} + senh\frac{\beta}{k}\,senh\frac{\gamma}{k}cosA$$

In the limit when $k \to \infty$, (using the exponential form of hyperbolic functions) this equation becomes: $= +\alpha^2\beta^2\,\gamma^2\,-2\,\beta\gamma\,cosA$, which is the fundamental formula of Euclidean plane trigonometry (cosine law).

Using another formula from spherical trigonometry involving the angles of a triangle, given by:

$$cos\hat{A} = -cos\hat{B}cos\hat{C} + sen\hat{B}\,sen\hat{C}\,cos\frac{\alpha}{k}$$

and replacing by the imaginary radius, using hyperbolic functions, it is:

$$cos\hat{A} = -cos\hat{B}cos\hat{C} + sen\hat{B}\,sen\hat{C}\,cosh\frac{\alpha}{k}$$

In the special case when $\hat{A} = 0$ and $\hat{C} = 90º$ we have:

$$cosh\frac{\alpha}{k} = \frac{1}{sen\hat{B}} \quad (*)$$

Equation (*) describes the triangle in which one angle is equal to zero and the two sides containing that angle would be infinite and parallel (asymptotic), given by figure 2.5, where sides AC and AB are parallel, considering that two parallels are not Euclidean straight lines.

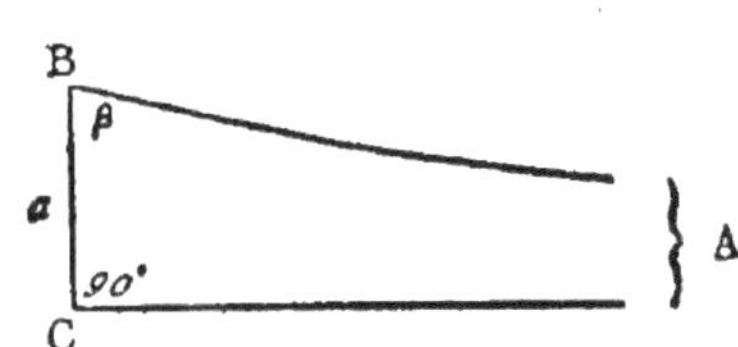

Figure 2.5: Taurus triangle, where the sides AB and AC are infinite and asymptotic

Another property that can be obtained from spherical trigonometry considering the imaginary radius is the relation between the area of a triangle and its defect, as predicted by Lambert (BONOLA, 1955, p. 81). Even though it obtained an answer to the problem of the sum of the angles of the triangle, Eq. 2.1 assumes a sphere of imaginary radius which has no geometrical meaning, and answers only the acute angle hypothesis. In the case of the obtuse angle hypothesis, spherical geometry has no imaginary radius, but should have different definitions than Euclid uses in plane geometry. The work of Taurinus can be considered as the first to deal with a non-Euclidean geometry (HOUZEL, 1992, p. 7).

2.4 - BOLYAI AND LOBACHEVSKY

During the period when Gauss was publishing his work on differential geometry, two mathematicians were developing, independently of each other, their theories involving the postulate of parallels, following Euclid's axiomatic model: Janos Bolyai and Nikolai Lobachevsky.

Janos (or Johann) Bolyai (1802-1860), Hungarian, was the son of Wolfgang (or Farkas) Bolyai (1775-1856), friend of Gauss. Wolfgang had tried to prove the fifth postulate in two works, but what he did was to find similar forms of it (BONOLA, 1955, p.60).

Wolfgang's demonstration caught the attention of his son, Janos, who constructed what he called the *Absolute Theory of Space*. Initially Janos intended to follow the same path as Saccheri and Lambert, but due to his father's mistakes he ended up following the axiomatic method of the Greeks, without deciding, *a priori, on* the validity or otherwise of the fifth postulate. With the collaboration of Szász, his friend at the *Royal School of Engineering*, Janos had the first ideas regarding parallelism (BONOLA, 1955, p. 97). Both Janos as Lobachevsky began their theories from the idea that *more than one parallel to a given straight line could pass through the same point*, which contradicts one of the consequences of the fifth postulate of Euclid. In a manuscript sent to his father in 1823, Janos wrote that he was about to finish a work on the theory of parallels in which he had made great discoveries.

In 1829, Janos sent the final paper to his father, who published it as an appendix to his book in 1832 (*Tentamen*). In the paper16 Janos provides a new definition for parallel, which depends on the "angle of parallelism":

> *"If the ray AM is not cut by the ray BN, situated in the same plane, but is cut by any other ray BP comprised in the angle ABN, we shall call the ray BN parallel to the ray AM; this is designated by BN || AM." (BOLYAI, [1832] 1955, p. 5).*

If we look at figure 2.6, we see that the definition of parallel used by Janos Bolyai does not imply equidistance between parallel rays, and so we can imagine the ray BN as being asymptotic to AM, while still being parallel.

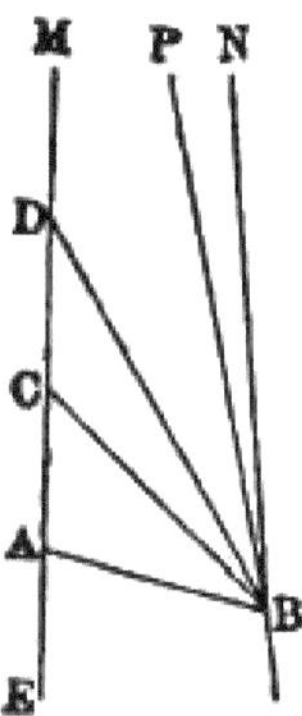

Figure 2.6: According to Bolyai's definition, the radius BN is parallel to the radius
AM.

Based on the properties of parallels as defined by him, Bolyai proved a series of theorems for the "new" geometry, all independent of the fifth postulate. One of these is the one concerning the sides of a triangle:

> *In any spherical triangle, the sines of the sides are like the sines of the opposite angles (BOLYAI, [1832] 1955, p. 21).*

Some of the most important results found by Bolyai in the construction of the *absolute theory of space* were the definition of parallels and their properties independently of Euclid's postulate, and the deduction of a geometry independent of the fifth postulate, where a parameter *i* appears, which he calls the *absolute unit of length*. When *i* tends to infinity, the geometry approaches the Euclidean geometry.

Although he did not start from the same initial hypothesis as Taurinus, that is, the hypothesis of the acute angle, Bolyai developed a very similar theory, which departs from the formulas of plane trigonometry to construct another trigonometry that depends on a parameter, which in Taurinus' case corresponded to an imaginary geometry.

In the same period that Janos was researching the postulate of parallels, Nikolai Ivanovitsch Lobachevsky (1793-1856) was, unaware of Bolyai's work, conducting research on the fifth postulate at Kazan University. Lobachevsky's first paper on the fifth postulate was published in 1829 and contained the basic principles of a new geometry (called *Imaginary Geometry* or *Pangeometry*) in which more than one parallel to a given line can pass through a point, which contradicted the Euclidean postulate and was the same as Bolyai's hypothesis. He published his theory before Bolyai (in 1829).

Lobachevsky published his work, *On the principles of geometry*,17 in a journal of *Kazan University*, in Russian, which did not make it known (GRAY, 1989, p. 106). In the period from 1835 to 1840, Lobachevsky published 5 papers, in other languages, in which he discussed the principles and applications of imaginary geometry and its relation to the postulate of parallels.

Unlike Bolyai, who remains indifferent to the fifth postulate and thus developed his theory, Lobachevsky has the impossibility of demonstrating this postulate as the main reasons for developing his theory as are mentioned in the introduction of the work Geometric Research on the Theory of Parallels:

> *In geometry I find certain imperfections which I believe to be the reason why this science, apart from the transition to analysis, may not yet have made advances from the state it has been in since Euclid. As part of these imperfections I consider the obscurity in the fundamental concepts of geometrical magnitudes, and finally the very important gap in the theory of parallels, in filling which all the efforts of mathematicians have been in vain. (LOBACHEVSKY, [1840] 1955, p. 11)*

Lobachevsky constructs a new geometry based on different definitions than Euclidean geometry, and assigns different properties to straight lines, such as (LOBACHEVSKY,[1840] 1955, pp.12-13):

1. A straight line adjusts about itself in every position. By this I mean that during rotation of a surface containing it, the straight line does not change its place if it passes through two fixed points on the surface (by this I mean that if we rotate the surface containing the straight line through two points on the line, the line does not move).

2. Two straight lines cannot intersect at two points.

3. If a line lies in a plane bounded by a boundary, then that line when it is

long enough on both sides, it must cross the boundary and thus divide the plane into two parts.

4. Two straight lines perpendicular to a third never cut each other, no matter how long they are extended.

The definition that Lobachevsky gives for straight lines makes it clear that he is not thinking of curves, showing that his proposal is for a new geometry in which straight lines have properties different from those of Euclidean geometry. Among other definitions that Lobachevsky provides, he includes that of parallel lines as (LOBACHEVSKY, [1840] 1955, p.13):

> *16. All straight lines in a plane starting from the same point may, with respect to a given straight line in the same plane, be divided into two classes: secant and nonsecant. The lines bordering this and that class of lines will be called parallel to the given line.*

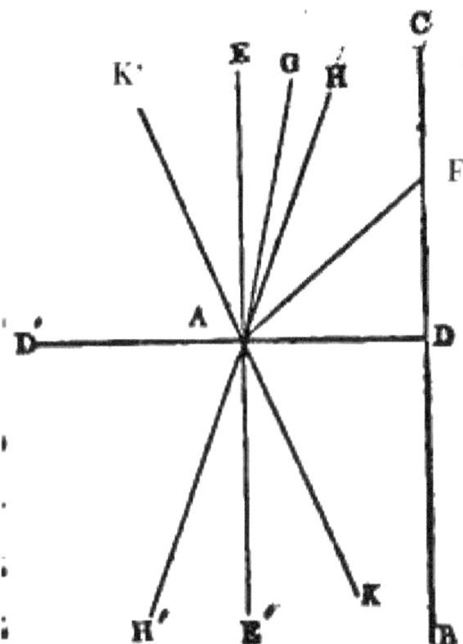

Figure 2.7: In Lobachevsky's definition of parallels, AF is secant to the DC line, AG is nonsecant to the DC line and AH is parallel to the DC line

Let AD be perpendicular to BC and AE (figure 2.12). Some lines from the right angle EAD meet the line DC, such as AF; others do not meet the line DC, such as the perpendicular AE. Assume the possibility that there are other lines that do not cut through DC, such as AG, however extended they may be. Between the secant lines, like AF, and the non-secant lines, like AG, there are lines like AH, parallel to DC, which are borderlines, since on one side of AH, all lines cut DC, like AF; and on the other side of AH, all lines do not cut DC, like AG. The angle HAD between parallel AH and perpendicular AD is called the *angle of parallelism*, denoted by P(p), where p is the distance AD.

If P(p)=p/2, AE can be extended, obtaining AE', which will also be parallel to the extension of the line DC, which is DB. This case corresponds to Euclidean geometry, which Lobachevsky calls ordinary, where the distance between the parallels is constant. But if the angle of parallelism is less than *p/2*, we have two parallel lines: AH parallel to DC and the line AK,

parallel to the extension DB of the line DC, which defines two sides of parallelism: the side where AH is and the side where AK is.

With this definition of parallels, Lobachevsky established new propositions, such as (LOBACHEVSKY, [1840] 1955, pp. 17-45):

18. Two lines are always mutually parallel.

19. In a right-angled triangle the sum of the three angles cannot be greater than two right angles.

24. The more two parallel lines are extended on the parallelism side, the closer they will be to each other.

31. We call a boundary line (horocircle) the curve on a plane for which all perpendiculars erected at the midpoints of the chords are parallel to each other.

In Figure 2.8, the line ACH is the horocircle, the segment AC is the chord, from which, at the midpoint, D, comes the perpendicular DE, which will be parallel to the line FG, which is perpendicular to the midpoint of the chord AH.

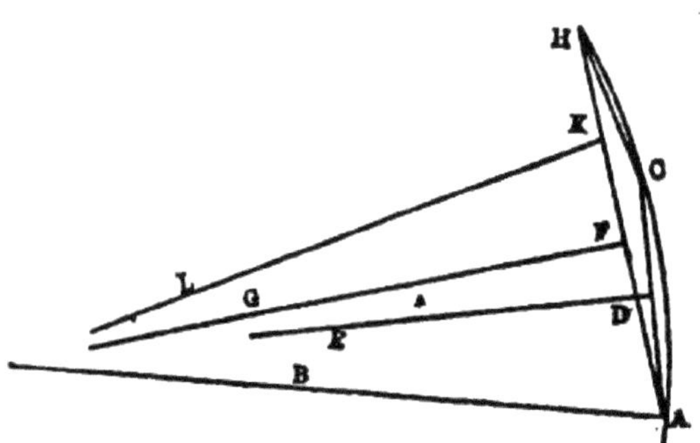

Figure 2.8: Arc AH forms the horocircle, where lines KL, FG, Ab and DE are parallel

34. Boundary surface (horosphere) is the surface formed from the revolution of the boundary lines around one of the axes. In Figure 2.9, W represents the horosphere.

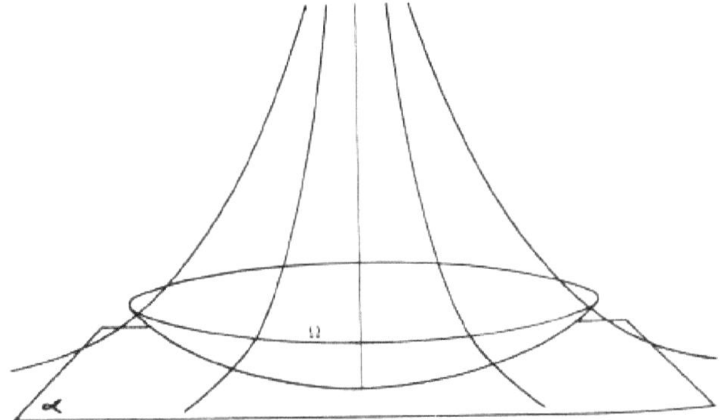

Figure 2.9: The Ω represents the horosphere

Lobachevsky also succeeded in finding, in the case of the spherical triangle, the relation between the defect in the sum of the angles of the triangle and the area of it on the sphere (LOBACHEVSKY, [1840] 1955, pp. 24-25).

Pangeometry is then founded on a series of propositions and the negation of the fifth postulate, where parallels are not equidistant but asymptotic, and allows Lobachevsky to construct a trigonometric formula using the angle of parallelism, to determine the relationship between the sides and angles of a right triangle. In this case, let a, b and c be the sides of the triangle and A, B and C be the angles opposite the respective sides. Thus Lobachevsky deduces the equations relating these angles and sides as (LOBACHEVSKY, [1840] 1955, pp. 41-44).

$$sen\, A \tan \Pi(a) = sen\, B \tan \Pi(h),$$

$$\cos A\ \cos\Pi(b)\cos\Pi(c) + \frac{sen\Pi(b)sen\Pi(c)}{sen\Pi(a)} = 1,$$

$$cotang\, A\ sen\, C\ sen\Pi(b) + \cos C = \frac{cos\Pi(b)}{cos\Pi(a)},$$

$$\cos A + cosBcosC = \frac{senBsenC}{sen\Pi(a)}.$$

$(**)$

Assuming the sides of the triangle to be very small and using one of Lobachevsky's propositions that relates the angles and sides of a right triangle22 (LOBACHEVSKY, [1840] 1955, p. 39), equations (**) are the equations of Euclidean geometry that allow finding that the sum of the angles of the triangle can be less than two right angles in pangeometry.

The (**) are very similar to the one found by Taurinus, and can be written, using the notation of hyperbolic trigonometric functions (i.e., assuming the sides as $a\sqrt{-1}$, $b\sqrt{-1}$ and $c\sqrt{-1}$ of a spherical triangle), as

$$\frac{senA}{sena} = \frac{senB}{senb},$$

$$\cos A \, tangh \, b \, tang \, c + \frac{\cosh a}{\cosh b \, \cosh c} = 1$$

which can be reduced to:

$$cosA \, senh \, b \, sen \, c + \cosh a = \cosh b \, \cosh c \quad (***)$$

We can see that Equation (***) is very similar to Equation (**), making the appropriate change of parameters, but there is no evidence that Lobachevsky or Bolyai were aware of Taurinus' work (GRAY, 1989, p.117).

The works of Lobachevsky and Bolyai are alike in many points. One of them is the definition of parallel lines from a bundle of lines, which serves to establish and prove several theorems.

A point that differs in the works is the goal set. For Bolyai, geometry could be restricted only to a "mental exercise" while for Lobachevsky there is a need to empirically test the validity of the new geometry (CICENIA, 1994, p. 13).

Moreover, he also believed that concepts do not exist, but are "sensed", a position that will be the same adopted by Poincaré and Helmholtz, as we shall see in the next chapter.

> *Surfaces, lines and points, as defined in geometry, exist only in our representation; since we actually measure surfaces and lines through our bodies. [In fact, we know nothing in nature except motion, without which our impressions are impossible. Consequently all other concepts, for example, geometrical concepts, are artificially generated by our understanding, which is derived from the properties of motion; this is because space in and of itself does not exist for us (LOBACHEVSKY, apud TORRETTI, 1984, p.65-82).*

Besides contradicting the conception of space accepted until then, with different properties for "straight lines", the existence of a new geometry also questioned the conception of geometry as an empirical science. We can speak of an abstract mathematical space, where properties are defined from a set of theorems and axioms; and a physical space that has its properties admitted from observations and measurements. Geometry as an empirical science implied its application to physical space, in addition to its formulation for mathematical space.

Several authors, except Lobachevsky, were concerned only with mathematical space. The space proposed by Newton's theory and Kant's philosophy, was basically Euclidean and would also be equivalent to physical space (JAMMER, 1970, pp. 118-147). Lobachevsky believed that physical space could be represented by a mathematical space in which non-Euclidean geometry would be valid.

> *The futilities of all these efforts of the last two thousand years since the time of Euclid make me suppose that in geometry the concepts themselves do not imply the truth whose proof we seek and whose justification, like the justifications of other natural laws, can be reached through experience, as for example by astronomical observations (LOBACHEVSKY, apud JAMMER, 1970, p. 149).*

Therefore, Lobachevsky's geometry could have examples in which it was the only possibility of explanation, which he supposed would happen in large-scale experiments, such as astronomical measurements. Only with astronomical measurements was there any hope of applying pangeometry to the explanation of physical space.

> *So imaginary geometry becomes ordinary geometry when we suppose that the sides of a right triangle are very small.*
>
> *I have, in the scientific bulletins of Kazan University, published some research regarding the measurement of curved lines, of plane figures, of the surfaces and volumes of solids, as well as regarding the application of imaginary geometry to analysis. Equations 8 [2.4] have in themselves sufficient grounds to consider the hypothesis of imaginary geometry as possible. Thus, there is no other means than astronomical observations, to judge accurately the calculations which belong to imaginary geometry. (LOBACHEVSKY, [1840] 1955, p. 44).*

Lobachevsky intended, through analysis of astronomical data, to determine whether, in this case, the theory of imaginary geometry would apply. To do this he applied the theory of pangeometry to determine the parallax of a star and compare this value with that found empirically. He discussed the case of a fixed star C with parallax $2p$, located exactly on a point A of the Earth's orbit (relative to the plane of the orbit). At the opposite point B of the orbit, whose distance *AB=a*, the parallel *AC* would make a parallelism angle P*(a)* (figure 2.10). If the line *BC* is parallel to the line *AC*, then π*(a)* ³ ≥ *(p/2) - 2p*. From the calculation of the angle of parallelism, he found that *a < tan 2p*. Considering the smallest parallax that had ever been found (corresponding to the star Sirius, which was 1.24") and estimated that *a<6.012.10⁻⁶* , where *a* has dimension equal to 1. Lobachevsky also calculated the defect in a solar system triangle formed by a base and a height approximately equal to the diameter of the earth's orbit, finding the value of 0",000003727 (LOBACEVSKY, 1898, pp. 19-24).

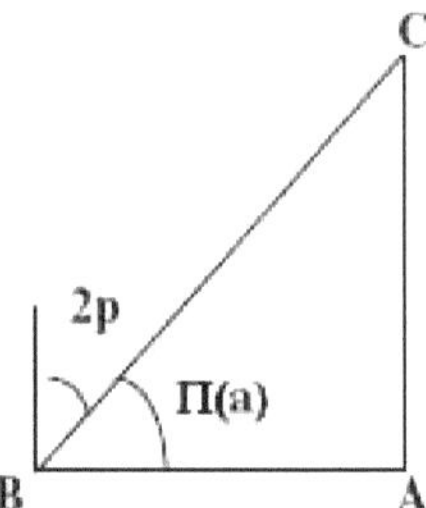

Figure 3: AB is the diameter of the Earth and C is the position of the fixed star with parallax
2p

Another difference between Lobachevsky's and Bolyai's works was the use of Pangeometry for the calculation of definite integrals, which shows that he was interested in the application of non-Euclidean geometry and not just its axiomatisation:

> *From what has been seen, two expressions for the value of the area of a closed polygon can be deduced, one expressed by a definite integral, the other dependent only on the measure of the angle of a polygon. The two calculated values must be equal. In this way a new method of finding the value of many definite integrals is obtained, values which would be difficult to solve otherwise (LOBACHEVSKY, 1867, p. 314).*

The angular defect of a triangle or a polygon is proportional to its area in pangeometry, and since it is possible to obtain the angles of a triangle in pangeometry using trigonometric relations involving the angle of parallelism (Eqs. 2.4), it is also possible to obtain the area of a polygon using trigonometric relations between its angles. Therefore, it is possible to obtain the area of a triangle, or a polygon, without calculating an integral. Lobachevsky also proposed some applications to mechanics, relating velocity and distance with angles and lines, but they were not widely disseminated (DANIELS, 1975, p. 80).

The works of Lobachevsky and Bolyai were not very well received at first, and they remained practically isolated for a long time. While Bolyai published only the appendix to his father's work, Lobachevsky published several works on the subject, also translated into

French and German, such as: *Geometria Imaginaria*, also translated into French in 1836; *Pesquisas Geométricas sobre a teoria das paralelas*, in German in 1840;

Pangeometria, with a translation also into French in 1856; plus two more papers written only in Russian, published in a journal of *Kazan University*.

The different conceptions of space that have emerged since Plato up to the current concept have had a great influence on the acceptance of non-Euclidean geometry. Finite or infinite space, nature of space, *n-dimensional* space, etc., are several issues that for years have served as a basis, as well as helped clarify the properties of non-Euclidean geometry and its application in mechanics, but are not part of the main objective of this work.

2.5 - GAUSS GEOMETRY

Carl Friedrich Gauss (1777-1855) developed work in various fields, involving astronomy, mathematics, physics, etc. In the period from 1818 to 1825, he made a series of field measurements for the King of Hanover (figure 3.1). This fact increased his interest in the study of geodesics, which came from the study of mapping the earth's surface on a sphere and a plane in 1816 (MAY, 1981, p. 298).

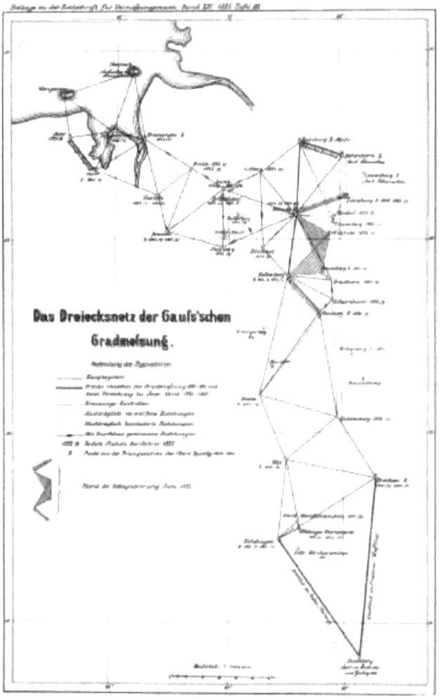

Figure 3.1: Gauss triangulation map of a region in Germany

With the end of the field measurements, Gauss devoted himself to calculations and analysis of the results, and in parallel, in October 1827, he presented to the Royal Society of Göttingen the work *Disquisitiones generales circa superficies curvas,* which became his fundamental work introducing differential geometry to any curved surfaces. In it, Gauss treated the curvature of a curved surface differently from what had been known since Euler. Besides relating it to the corresponding area on an auxiliary sphere, Gauss found equations for the curvature that depended on the intrinsic coordinates of the surface, which allowed him to define his *Theorema Egregium,* and to construct a language in differential geometry that would serve as the basis for Riemann's later development of a non-Euclidean geometry (1854).

2.6 - THE NATURE OF THE CURVED SURFACE

Gauss begins his work by defining an auxiliary sphere for the study of any curved surface. The auxiliary sphere has a unit radius, with no dimensional unit. Gauss uses the auxiliary sphere to describe directions of lines and planes. The direction of any line can be associated with a radius parallel to the sphere (Figure 3.3), and this radius can be represented by the point where it touches the surface of the sphere. Thus the direction of a line can be represented by a point on the surface of the sphere (GAUSS, 1827, Sections 1 and 2). In Figure 3.2, the direction of the line *r* is given by the point P on the auxiliary sphere.

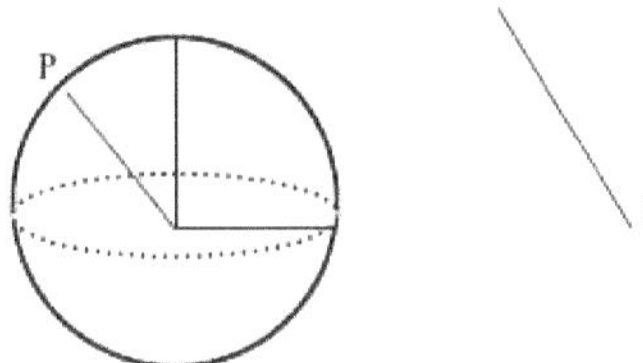

Figure 3.2: Point P on the auxiliary sphere represents the direction of the straight line r, parallel to the radius of the auxiliary sphere.

Note that, strictly speaking, we always have *two* opposite rays on the sphere, parallel to any straight line. Gauss considers always *oriented* straight lines, which eliminates the indeterminacy of this correspondence.

Let us consider a differentiable curved surface: it is possible to draw a tangent plane at every point on the surface and, at this point, a normal to the tangent plane. The normal, or what we could describe as a normal vector, has its direction defined by a corresponding (parallel) radius on the auxiliary sphere and by a point on the surface of the sphere. In this way, each point on the curved surface is associated with a point on the surface of the reference sphere. Since on the curved surface the normals at different points may have the same direction, the corresponding rays may overlap. In the same way, a figure on the curved surface has several associated radii on the auxiliary sphere and thus forms a figure on the auxiliary sphere as well. The normal to a point $P=(x,y,z)$ on the curved surface is represented by the point P' on the surface of the reference sphere, which has coordinates X, Y, Z. Since the radius of the reference sphere is one, the coordinates of this point are the cosines of the angles formed by the normal with the x, y, z directions.

The direction of a plane can also be characterised using the sphere as a reference. Consider any plane, and a plane parallel to it passing through the centre of the sphere and cutting it into a great circle. This great circle represents the direction of the first plane (Figure 3.3). However, the direction of the plane can also be represented by the direction of its normal. Since the normal to the plane can be represented by a point on the reference sphere, then the direction of the plane can be described by this point.

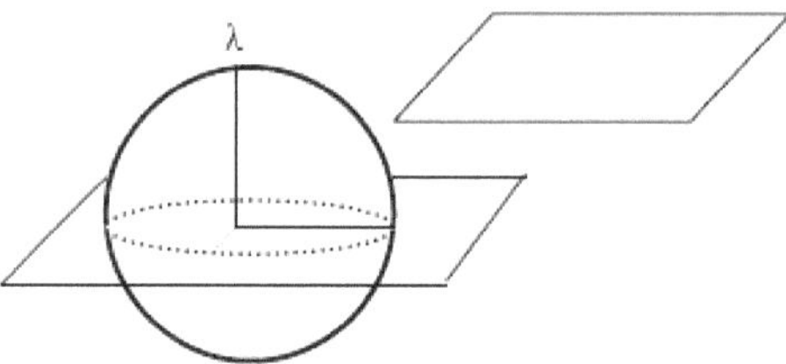

Figure 3.3: The point λ on the auxiliary sphere represents the direction of the plane, parallel to the plane that cuts the sphere forming a great circle

With the association of any line, plane and angles to points on the surface of the auxiliary sphere, the study of any curved surface, which can always be determined point by point through tangent planes, and normal to these planes, comes to be done through spherical trigonometry, which by the beginning of the 19th century was already well known.

Once the auxiliary sphere is defined, it is possible to study the nature of the curved surface by analyzing the normal to the tangent plane at each point on the curved surface, using for this the coordinates of each point and its correspondent on the auxiliary sphere. To this end, Gauss adopts three methods of describing the surface, always assuming that it is a "smooth and continuous" curvature, that is, in current language, in which the first and second order derivatives do not cancel each other out (GAUSS, 1827, sections 3, 4 and 5). We will address only the first method, which is the simplest.

The first method consists of defining a function describing the surface as $W=f(x,y,z)=0$, where x, y, z are the Cartesian coordinates of each point on the surface. The differential form of this relation is written by Gauss as $dW=Pdx+Qdy+Rdz$, where the coefficients P, Q, R are of course the partial derivatives of W *with* respect to each of the coordinates. In this and other points, the notation used by Gauss is different from the current one, which makes it somewhat difficult to understand his reasoning.

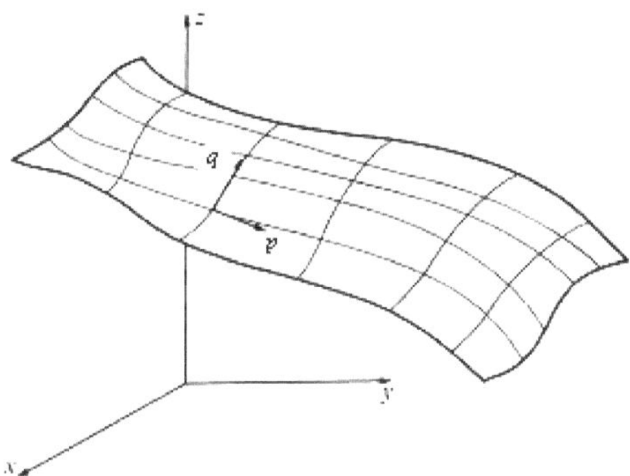

Figure 3.4: The coordinates *p* and *q lie* on the curved surface

2.7 - RIEMANN MULTIPLICITY

Georg Friedrich Bernhard Riemann (1826-1866) was born in Hanover in 1826, the second of six children. He had his father as a teacher until the age of 10, and under his influence he entered the University of Göttingen in 1846 for theology, changing the course to philosophy after becoming interested in mathematics. In 1847 he moved to the University of Berlin to study with Steiner, Jacobi, Dirichlet and Eisenstein, returning to Göttingen in 1849 where he graduated and had his habilitation thesis, *Ueber die Hypothesen, welche der Geometrie zu Grunde liegen,* presented in 1854 (FREUDENTHAL, 1981, p. 456). Riemann does not have a large quantity of works, but those he produced influenced both mathematics and physics, especially Einstein's theory of relativity.

In this paper, an introduction is made to the concept of *n-dimensional* multiplicity and generalization of the concept of curvature given by Gauss in *Disquisitiones generales circa superficies* to an *n-dimensional* surface. Besides the introduction of new mathematical concepts, there is also the introduction of a new idea about the concept of space itself.

Unlike the position of most mathematicians and philosophers of the time, who admitted the metric structure as fixed and independent of the physical phenomenon to which it is associated, Riemann considers that space is devoid of form, and acquires it only through the material which fills it and determines its metric relations. Later, the conception that the space can have its metric variable and "influenced" by the phenomena led some authors to associate the metric to a *metric field*, analogously to the electric field, magnetic field, etc., where the metric properties would alter the phenomena (SCRIMIERI, 1992, pp. 155-156).

Riemann space is different from Euclidean space. Euclidean space is associated with geometry based on the *Elements*. Riemann space is amenable to any metric relation, of which Euclidean space is one of the cases in which the dimension is equal to 3, and which has its properties established by experience32 (TORRETI, 1984, p. 84).

35

On the hypotheses... had little repercussion and only got important reactions from other mathematicians after Riemann's death. The first publication of his work came only in 1867, in German, when Dedekind published Riemann's complete works. The first known English translation was made by Clifford and published in 1873. Interested in Riemann's work and responsible for editing his memoirs, together with Weber, Dedekind wrote a paper in which he fully investigated Riemann's thesis. Beltrami, who had been working with differential geometry, mainly with geodesics, after becoming acquainted with Riemann's work, begins to apply it to Lamé's theory of differential parameters, which was the starting point for Ricci and Levi-Civita to develop the mathematical tools of the absolute differential calculus. Following Riemann's work *On the hypothesis...*, in the next topics we will present the definition of multiplicity, the metric properties of multiplicity and how this new concept is applied in space.

2.8 - DEFINITION OF AN N-DIMENSIONAL MULTIPLICITY

For Riemann, the geometry developed from Euclid to Legendre did not clarify certain axioms because it did not assume the existence of a *Mannigfaltigkeit*. The term introduced by Riemann has different translations. In the English translation, *manifoldness*; and in the Portuguese translation, it is more common to find it as *variety*. However, we have chosen to use the literal translation of the German word, that is, *multiplicity*.

He does not explicitly mention the fifth postulate, but considering that Legendre's work *Éléments de géométrie* was widely known at the time, we can assume that this is the postulate he was referring to, since Legendre's work contains a theorem on the sum of the angles of a triangle that does not depend on the postulate of parallels.

A multiplicity is defined as an extension containing points or elements, which in the first case can be continuous, and in the second case, discrete. A multiplicity would then be a mathematical "entity", formed by points or elements, which can have *n-dimensions*.

Riemann calls the constituents of multiplicity "specializations" (RIEMANN, 1959, p. 413). The simplest way to think of a multiplicity is to associate it with geometry, but other forms can be multiplicities. Euclidean space is a space where its points form a three-dimensional multiplicity. The set of all ellipses that can be determined individually from their major and minor semi-axes forms a two-dimensional multiplicity. The possible positions of a mechanical system with *n* degrees of freedom constitute, in general, an *n-dimensional multiplicity,* since they are given by *n* equations (WEYL, 1952, p. 84).

In general, the characteristic of an *n-dimensional* multiplicity is that each point or element that forms it can be determined using *n* parameters, where *n* can be a finite number, but can also be infinite of various kinds.

> *There are multiplicities in which the determination of position requires not a finite number but an infinite series or a continuous multiplicity of determinations of quantity. Such multiplicities are, for example, the possible determinations of a function for a given region, the possible shapes of a solid figure, etc. (RIEMANN, 1929, p. 415).*

What defines the dimension of a multiplicity is the number of parameters needed to identify each point or each element on the multiplicity, whether discrete or continuous. If in a multiplicity there are two points where the mode of going from one to the other is unique, then this mode defines a linearly extended multiplicity, that is, one-dimensional, because with only one parameter it is possible to locate a point on the multiplicity. If there are two lines, then two parameters are needed to locate the point, because it can be on either line. If two parameters are needed, then two lines define a doubly extended multiplicity, that is, two-dimensional; and so on for the other dimensions, whether the multiplicity is discrete or continuous (RIEMANN, 1959, p. 414).

If a set of specializations, i.e. points and elements, forms regions in the multiplicity, which are defined by a mark or contour, then this set is called a Quantum. The comparison between the quanta and the multiplicity is determined by a process of counting (how many quanta there are in the multiplicity), if the multiplicity is discrete; or by a process of measurement, which can be done using a standard, or by comparison with the whole, if the multiplicity is continuous (RIEMANN, 1959, p. 413). Once the concept of an *n-dimensional multiplicity is* formed, the concept of place determination is reduced to the concept of quantity determination, that is, the determination of a point in a multiplicity from the determination of how many regions there are between this point and a given origin.

With the above definitions Riemann characterized a multiplicity, its elements and their transformation and the way of locating a point on it by a quantity and not by position, that is, how many coordinates determine the location of the point. This last condition can be considered Riemann's first hypothesis, that is, that an element of the multiplicity is determined by n independent continuous quantities $x_1, x_2, ...x_n$ and, in a reciprocal way, n quantities determine a single element of the multiplicity (FERNANDES, 1989, p. 97).

According to Riemann, space can be unlimited even though it is finite. Unlimitedness is bound to extension relations while infinitude is bound to metric relations. The best example of this is a sphere which is unlimited, that is, an object can be moved all over the sphere without encountering a barrier or a point that cannot be surpassed (no boundaries); but it is finite, because no matter how many turns around the sphere the object always returns to the same place. To the infinitely small he relates the experiments with atoms, and to the infinitely large, the astronomical experiments.

Extension relations, metric relations and the hypothesis about a non-zero curvature, that is, a non-Euclidean geometry, can be proved and are connected with another science, physics, which Riemann claims is not part of the present discussion. Worthy of note is Riemann's remark that "*metric or extension relations determine how space must* be".

Riemann's ideas were very different from the proposed differential geometry and *n-dimensional geometry*, and his presentation, as in most of his works, was much more directed toward the philosophical aspects of the new view of geometry than toward calculations or applications. The application of the theory presented was only partially made in 1861, in a paper in which he analyses the conduction of heat.

Although Riemann also discusses surfaces, he goes beyond what Gauss did, for in dealing with three-dimensional space, or space with more dimensions, he generalizes the ideas to a generic curved space, while Gauss remains restricted to Euclidean space. Despite dealing with a different space, Riemann's way of approach is completely different from that of Lobachevsky and Bolyai. Riemann is not concerned with the question of parallels or the validity of postulates (axiomatic geometry), but with the definition and description of the properties of a new space, based on metric relations, which are studied using differential geometry.

2.9 - N-DIMENSIONAL SPACE MECHANICS

It is not possible to establish a milestone where the study of non-Euclidean geometry and mechanics come together, for it is clear from what has been seen so far that progress made in both fields after the introduction of differential calculus is inseparable.

If it was possible the description of a surface using its intrinsic properties, it could also be described the motion of a free particle on this surface using the same resource of the line element. The extension to n particles should take into account the interaction between them and actions that modify the motion of the centre of mass, but it would still be restricted to the motion on a two-dimensional surface, immersed in Euclidean space.

With the generalization of differential geometry to n variables, mechanics in a space with more than three dimensions became predictable, even if it was not possible to visualize it or to prove it through experiments. The development of the formalism that would describe motion became the difference between scholars of the time.

At the same time, Lipschitz and Christoffel developed two different methods for the analysis of motion: the first in a more applied way to the problem of mechanics; the second in a more analytical way; but neither of them was concerned with the principle of minimum action and its relation to geodesics. This relationship only appeared in the works of Darboux, and in a more comprehensible way than in the other two authors.

After Darboux's work, mechanics was again constructed only by formal methods, and this happened with the introduction of the absolute differential calculus by Ricci and Levi- Civita. Absolute differential calculus had its origins in Lamé's differential parameters, which in turn were related to the line element described with the intrinsic properties of surfaces and directly related, since Euler, with the description of the geodesic and maximization and minimization of quadratic functions.

3 - GEOMETRIC MODELLING, COMPUTER GRAPHICS AND FRACTALS

The themes Modeling and Manipulation of graphic objects are inserted in an area of Computer Graphics called *Geometric Modeling*. To better elucidate the results it is convenient to represent, in some graphical output, the object previously modeled. The area of computer graphics that transforms data into images is called *Image Synthesis.*

Computer graphics is the science that basically studies a set of techniques that transform data into an image presented through a graphic device. This data can be organised through a mathematical model whose geometry resembles that of the real-world object it is desired to model. In this way, Computer Graphics can be considered an area of Applied Mathematics.

Geometric modelling deals with the problem of describing and structuring geometric data on the computer, i.e. finding a mathematical model capable of representing the desired object. An example of Geometric Modelling is obtaining a function that approximates an unknown one through a method of interpolation of some already known sample points.

The Image Synthesis, or Visualisation, processes the data generated by the system in the geometric modelling and produces an image that can be viewed on a graphic output device. From the mapping of some points, distributed around a certain object, a mathematical model is obtained that represents it through the Geometric Modelling. With this model it is possible to present an image that represents the shape of the object. This last process is an example of Image Synthesis. Image Processing takes an image as input data, processes it and produces another image as output data. The techniques of this area can be used, for example, to improve the quality of an old or damaged image, producing another, less noisy image.

Today computer graphics is applied in several other areas of research that need somehow, explore one of the characteristics of this area that are as follows:

1. Allow you to view objects that are still in the design phase

2. Visualise objects that are beyond the range of human visual perception

3. Visualising objects that escape from three-dimensional reality.

Computer graphics applications fall into three main areas:

1. Computer Aided Design and Production

2. Data Visualisation and Motion Visualisation

3. Human Machine Interaction

The main objective of this chapter is to bring together a set of models capable of building a representation, at first three-dimensional, whose shape is similar to that of a real-world object. In many computational applications what is most important is the representation of the faces of the objects to be presented, i.e. a surface that models it.

To create this representation, in a first phase a rough model is created through some simple geometric primitive, or the meeting of these primitives. This is a geometric modelling process. To sophisticate this representation, one can select a point with the *mouse* and drag it, deforming it conveniently. This method is known as *Direct Manipulation*.

The modeling of solid objects, i.e., those of which are relevant to the information of its interior, is more complex, since it has particular characteristics about its interior, in addition to maintaining the properties of curves and surfaces present in the border of such an object. Another important fact is that to model them, the capture of information inside them requires special technologies, such as tomographs.

Thanks to this fact, no information about the interior of the objects will be analysed, but rather a model that represents their faces. Such studies become extremely motivating with a view to future applications. Important applications result from these theories in Mathematics, for example. Among them, we can represent graphs of functions and surfaces, where the visualization entails a deepening and incentive in the study of topology, Differential Geometry, Differential Integral Calculus, among other mathematical areas.

Fractal geometry in the last 20 years in all its concepts has become a central tool in many sciences, such as: geology, meteorology, among others. At the same time, fractals are of interest to graphic designers and filmmakers for their ability to create new forms and more realistic artificial worlds.

In Computer Graphics, fractals, among other things, are used to represent elements of Nature such as craters, planets, coastlines, lunar surfaces, plants, ripples in water, representation of clouds; they are also of great importance for creating special effects in films, such as the creation of the planet Genesis in the film Star Trek II.

It is important to know that fractal geometry can be used to describe various phenomena in nature, where traditional geometry cannot be used.

> *"Clouds are not spheres, mountains are not*
> *circles, a bark is not continuous and neither does*
> *lightning travel in a straight line"*
>
> Benoit Mandelbrot

Fractal geometry helps bring the natural sciences and computing closer to mathematical research, and is used as a major tool in various sciences such as geology and meteorology, among others.

Contributions of fractals, fractal geometry and its concepts:

- In biology - Study of the influence of the irregular surface of proteins on molecular iterations;

- In geography - Population growth models;

- In computing - Terrain and atmosphere generation with graphic modellers; image compression.

3.1 - APPLICATIONS OF FRACTALS IN COMPUTER GRAPHICS

Fractals help in the creation of new forms and more realistic artificial worlds, and in the representation of elements of nature that traditional geometry cannot represent.

3.2 - FRACTALS IN COMPUTING

Fractal dimensions have been used in:

- Geo-technologies for the classification and study of landscapes;

- Study of the behaviour of the World Wide Web (WWW);

- Mandelbrot sets are used in parallel applications such as benchmarks, cryptography, audio and video encoding and decoding.

The set of applications of these graphic objects still includes applications in medicine and in several other scientific areas that, somehow, need to visualise some element.

4 - GEOMETRIC MODELING OF FRACTALS IN COMPUTER GRAPHICS

4.1 - ORGANIZATION OF THE THESIS

After performing a bibliographic investigation on the subject, some possible approaches to modeling objects of non-Euclidean geometric order, especially from nature, will be presented, showing a classification already existing in the literature. Next we will divide modelling into two categories: 2D and 3D and briefly discuss the techniques that will be used in each one.

We will present the construction of procedural functions that will later be used in the modeling process. Before presenting the fractal functions, a short literature review on the subject is presented.

4.2 - RELEVANT WORKS IN COMPUTER GRAPHICS

Since the emergence of Computer Graphics, several works have been done to generate images from data, including the elements of nature. *Of Growth and Form,* by D'Arcy Thompson [Thompson 61] may be considered the first work on the subject. In this book, the author discusses geometry and the main physical laws that determine the shape of elements in nature. *Patterns in Nature* [Stevens 74] is a more recent work that discusses the shape of elements of nature and their similarities with some mathematical formulations. In the book *The Fractal Geometry of Nature* [Mandelbrot 82] Benoit Mandelbrot introduces the concept of fractals applied to natural phenomena, opening a vast field of research and development for this area, as will be seen in chapter 2 of this work.

Later Peitgen and Richter studied how to implement the theory of fractals through computational algorithms presented in the book *The Beauty of Fractals* [Peitgen et al 86]. In 1994 *Texture and Modeling - A Procedural Approach* [Ebert et al 94] was published, which contains a collection of results obtained by several authors, mainly in the area of procedural functions, giving much emphasis to elements of nature. Regarding the modeling of plants, a great reference is the book *The Algorithmic Beauty of Plants* [Prusiniewicz et al 90].

Other relevant works carried out at Puc-Rio, which can also serve as a great reference for this subject are *Textures Applied to the Radiosity Method* [Birtel 98], Texturing *Procedural Techniques [*Aguilar 98], *Techniques for Texture Generation in a Ray-Tracing System* [Faria 96] and *Perturbation and Textures in NURBS Surfaces* [Cavalcante 94].

Throughout this project some of these references will be discussed in more detail.

5 - FRACTALS

Just using the basic functions in a direct way would already be possible to create an infinite number of textures and models. However, in nature, the elements have a wealth of detail which in many cases could only be reasonably modelled mathematically by means of fractals. Benoit Mandelbrot was one of the first to delve into this topic. [Mandelbrot 82] is an excellent reference on the subject.

A simple but adequate definition for fractals is the one given by Kenton Musgrave in [Ebert et al 94, pp. 251]: "geometrically complex object, a complexity that arises by the repetition of a formula for variations of scale". In theory, to create a truly fractal object, this repetition would need to be infinite, for smaller and smaller scales.

In practice this infinite repetition is not necessary, since after a certain moment the differences generated by the recursions will be imperceptible on the media where the result is being sampled. This characteristic, on the other hand, is of great interest in the process of modelling an object or when creating a texture: the elements can be infinitely "enlarged" without the concern of losing resolution or details.

Thus, using the same formula one can, for example, generate a sky with several small clouds or a sky with a single large cloud rich in detail. By calculating only a finite number of iterations one is said to be limiting the bandwidth of the fractal function.

5.1 - WEIERSTRASS AND RIEMANN FUNCTIONS - CLASSICAL FRACTAL CONSTRUCTIONS

The first constructions and perhaps the most classical examples of known fractals were made in the last century and correspond to continuous curves that do not have a tangent at every point. These curves were constructed to show that a statement *obvious to* us, in fact was false:

> *"The graph of a continuous function has a well-defined tangent at all but perhaps a finite number of its points."*

The problem of the existence of a continuous function not differentiable at any point was proposed by Weierstrass by considering the function:

$$W(x) = \sum_{n=0}^{\infty} b^n \cos(a^n \pi x)$$

Where a is an odd integer, $0 < b < 1$ and $ab > 1 + \dfrac{3\pi}{2}$.

The following figures show graphs of the function $W(x)$ for $x \in 2\,[0\,,\,2\pi\,]$ and $n = 1;\,2;\,3;\,4$ and $n = 10$.

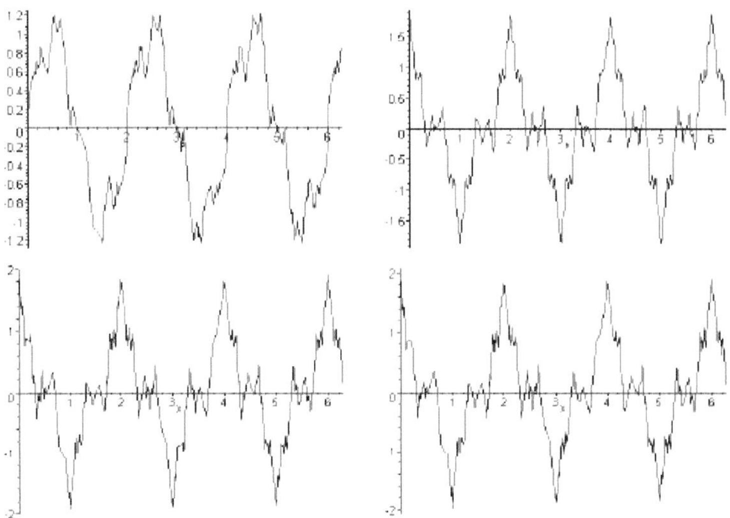

  The example proposed by Riemann to counter the previous statement is the following function, known today as the *Riemann function*.

$$R(X) = \sum_{n=1}^{\infty} \frac{sen\,(n^2\pi x)}{n^2}$$

The following figures show some graphs of $R(x)$ for $x \in [0,\,2\pi\,]$ and $n = 1;\,2;\,3$ and 10.

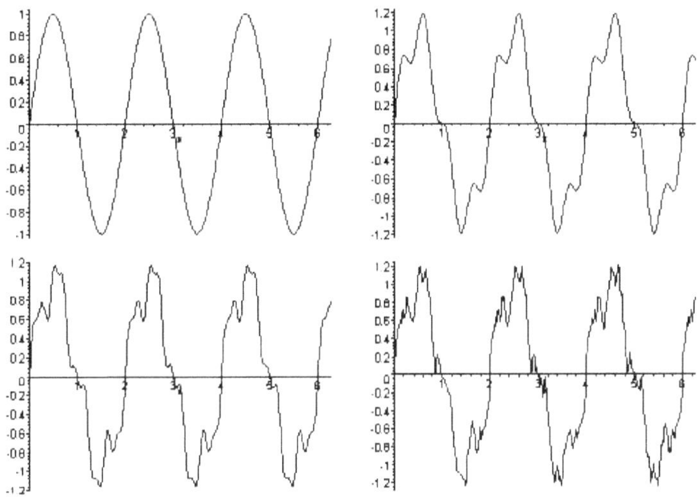

5.2 - SOME CLASSICAL EXAMPLES OF FRACTALS

5.2.1 - Bolzano curves

Here we saw that Bolzano in 1830 had already constructed a "geometrical" example of a continuous function whose graph had no tangent at every point. The construction is as follows. We take each segment of a polygon, Step 0 of the figure below. We divide each segment into two segments, Step 1. Then we divide, again, each segment in two, Step 2, so that we create more and more "angled" points.

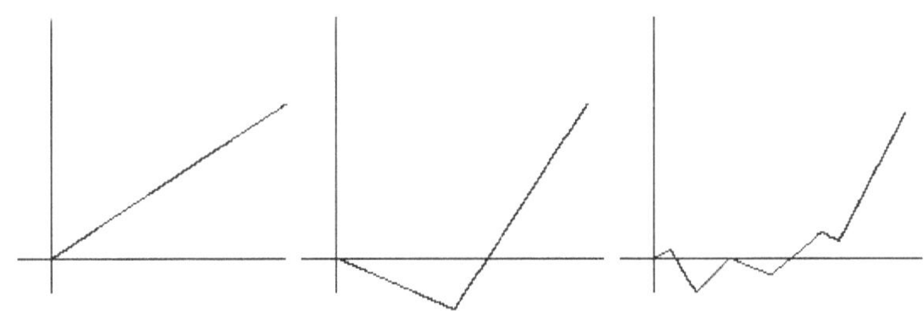

5.2.2   - Koch curves

H. Von Koch in 1904 constructed a continuous curve that has no tangent each of its points known as *Koch curve*. This curve is obtained from elementary form and also in steps. The procedure is as follows. We start with an interval, say [0; 1], dividing it into three equal parts, and replace the central part of the interval by two segments of length $\frac{1}{3}$. As shown in the figure below:

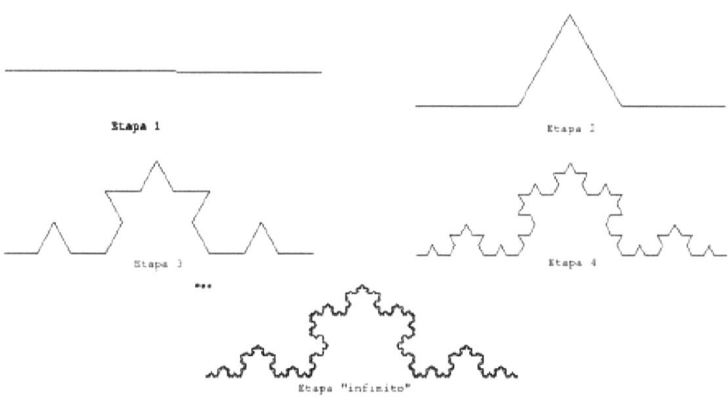

Figure 3.5 - Stages of the construction of Koch Curves. *Source* www.inf.ufsc.br/~visao/2000/fractals/image010.jpg

5.2.3   - Sierpinski triangle

This construction is similar to the previous constructions, but made from a triangle. Consider a triangular region in the plane, Phase 1 of the figure below. First, we divide the sides of the triangle at their midpoints, joining them to form four triangular regions, from which we eliminate the central triangular region, Phase 2. In Stage 3, we repeat the process of Stage 2, with each triangle obtained. And so on, successively, as shown in the figure.

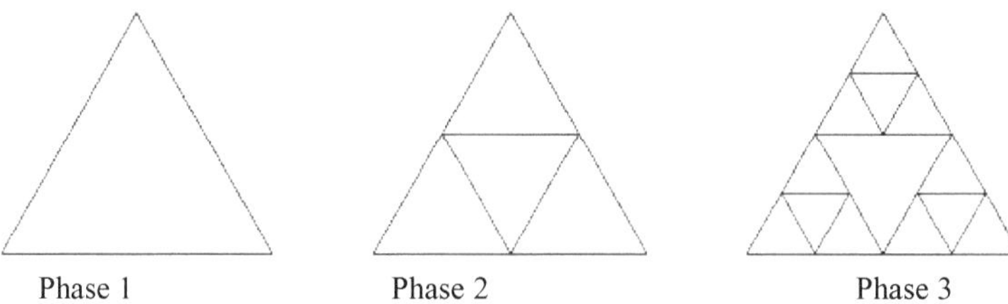

Phase 1   Phase 2   Phase 3

In the remaining triangular regions we repeat the previous process, and so on. This process is convergent and the figure is known as a *Sierpinski triangle*. The Sierpinski triangle is a "curve" which has infinite length and occupies virtually zero area in the plane. The "final" result is shown in the following figure.

Figure 3.6 - Sierpinski Triangle tending to Infinity - Source www.educ.fc.ul.pt

5.2.4   - Sierpinski curve

To obtain the Sierpinski curve, we start from an initial figure, which we call the generator obtained as follows:

- We initially consider an equilateral triangle. We fix one side as the base and mark the midpoints on the other two sides.

- We take only three segments. Two are supported by the sides of the equilateral triangle, starting at the base extremities and extending to the midpoints, and another segment joins the midpoints.

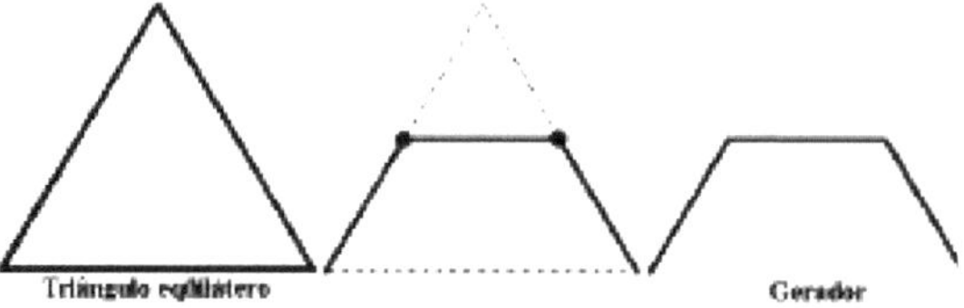

Figure 3.7 - Sierpinski Curve Generator

From the generator, we obtain the Sierpinski curve by repeating, indefinitely, the process described below:

We replace each segment of the generator with the figure of the previous level reduced by the factor $\frac{1}{2}$ conveniently as in Figure 3.7.

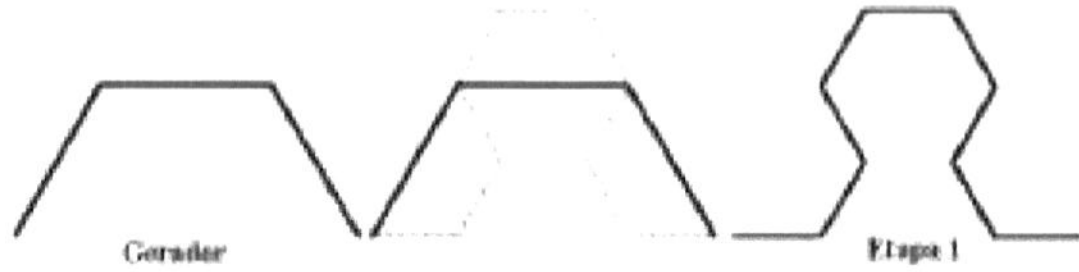

Figure 3.7 - Level 1 bend

In an algorithmic language we could describe the process of constructing the Sierpinski curve as follows:

Algorithm 1 - Sierpinski curve

Data: Generator (Figure 3.5)

k = 0

curve(k)=generator.

repeat indefinitely

curve(k + 1)=curve obtained by replacing, in an appropriate manner, each

generator segment by the reduced curve(k) of the factor $\frac{1}{2}$.

$k = k + 1.$

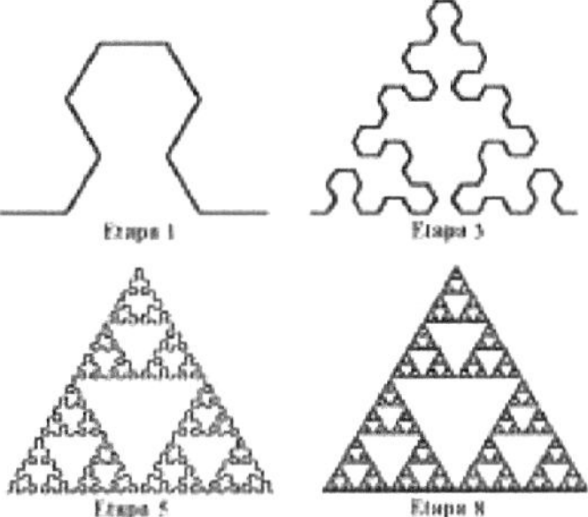

Figure 3.8: Different levels of obtaining the Sierpinski curve

Continuing the construction of the Sierpinski Curve, making k tend to infinity, will result in the Sierpinski triangle (Figure 3.6).

5.2.5   - Singer's ensemble

The Cantor set is formed by all the points $x \in [0; 1]$ such that in their representation in base 3 the coefficient 1 does not appear, that is, points of the form:

$$x = \sum_{j=1}^{\infty} k_j\, 3^{-j}$$

In the previous representation there are some special points, which correspond to the extreme points of the intervals formed through the process of construction of the Cantor Set, for example: $\frac{1}{3}, \frac{1}{9}, \frac{7}{9}$.

It is important to point out that there is a very important property inside the fractals theory and that will have relevance on the applications of this work: it is the fractal dimension. While in the Euclidean integer dimension the dimensions can be only integers (0 corresponds to a point, 1 to a line, 2 to a plane and 3 to space), in fractals one can have non integer dimensions. Thus, it is possible to change from one dimension to another in a continuous way. Graphically, one can think that this process corresponds to starting from a plane, for example, and gradually *filling in* parts of the 3D space, until it is completely formed.

6. MATHEMATICAL DEFINITION OF FRACTAL

6.1 - WHAT IS A FRACTAL?

The term fractal was coined by Benoit Mandelbrot, to designate a geometrical object that never loses its structure whatever the viewing distance. Fractal above all means self-similar. Mandelbrot classified his objects of study in this way, as they were fractal-sized.

Non-integer dimensions then became a way of quantifying qualities that would otherwise remain unquantifiable: the degree of irregularity or tortuosity of an object.

If you notice, all orthodox geometric shapes degenerate when they are enlarged or diminished. A circle on a much larger scale is nothing more than a straight line. Just bear in mind that only 500 years ago the Earth was thought to be flat. This, because at the human scale we see nothing more than a straight line on the horizon. However, most of the objects we deal with in our daily lives are not straight lines, spheres or cones.

Looking, for example, at the trunk of a tree, we can see that it is extremely rough and irregular. If we look at a small piece of it under the microscope we see new roughness and irregularities that we hadn't seen before. However, this image is very similar to the previous one. It is this regular irregularity that characterises a fractal.

6.2 - MATHEMATICAL DEFINITION

To begin, we can say that a Fractal is a "complicated" subset of a "simple" metric space, such as IR^2, IR, [0,1], C and $\hat{C}$, among others. Theoretically, any metric space, provided it is complete, can be used to construct a fractal, but in practice we are interested in those that can easily be represented graphically, as is the case of IR^2, IR, C, $\hat{C}$ and their subspaces. This "complicated" form is obtained as a result of a system of iterated functions (IFS). According to the approach I intend to take:

A Fractal is the fixed point of a system of iterated functions in a metric space with associated Hausdorf metric.

I will then explain each of the aspects necessary to understand the previous definition.

What is a metric space?

Any set X can be thought of as a space, where its elements are the "points" of that space.

To have a metric space, it is necessary to define a metric on X. A metric is an application d: X x X$\rightarrow$ IR that has the following properties:

- $d(x,y) = d(y,x), \forall\ x, y \in X$

- $0 < d(x,y) < ,\infty \forall$ x, y$\in$ X, x$\neq$ y

- d(x, x) = 0, x$\forall \in$ X

- d(x,y)$\leq$ d(x,z) + d(z,y),$\forall$ x, y, z$\in$ X

Having such an application defined on X, we say that X together with that application forms a **metric space (e. m.)** and we call it **(X, d)**.

What is a metric space with Hausdorf metric?

Since (X, d) is a **complete metric space** (that is, all its Cauchy succession $\{x_n\}_{n=1}^{\infty}$ on X has limit x$\in$ X), then **H(X)** represents the space whose points are the non-empty compact subsets of X. (A subset S of X is **compact** if all its succession has a subsucession with limit on S).

H(X) represents the space whose points are the non-empty compact subsets of X.

Let us now define a metric on H(X):

Let (X,d) be a complete metric space, x$\in$ X and A,B$\in$ H(X). To define the Hausdorf metric one must do so by going through the following steps:

    i.   Set Distance from x to A

   ii.   Set Distance from A to B

  iii.   Define Metric h(d) on $\mathcal{H}(X)$

    i.   We define distance from a point x of X to a compact subset B of X thus:

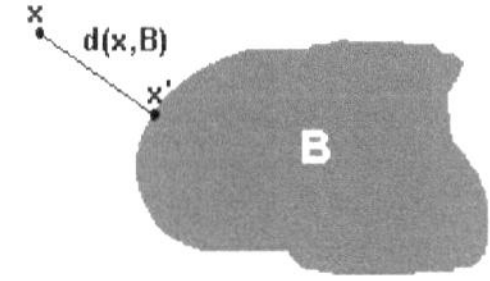

$$d(x,B) = min \{d(x,y) : y \ B\}\in$$

We know that this value exists due to the fact that B is compact and non-empty and so$\exists$ x'$\in$ B: d(x,B) = d(x,x').

ii. The **distance between two compact subsets A and B of X** is defined as follows:

$$d(A,B) = \max\{d(x,B) : x \in A\}$$

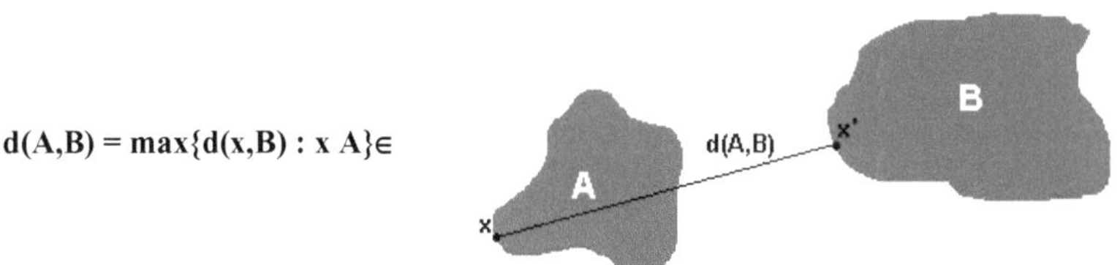

Again, due to the fact that A and B are compact and non-empty, we know that $\exists x \in A, \exists x' \in B : d(A,B) = d(x,x')$.

As one can easily see this definition of distance between subsets A and B of X cannot be regarded as a metric in $\mathcal{H}(X)$, since one does not always have $d(A,B) = d(B,A)$.

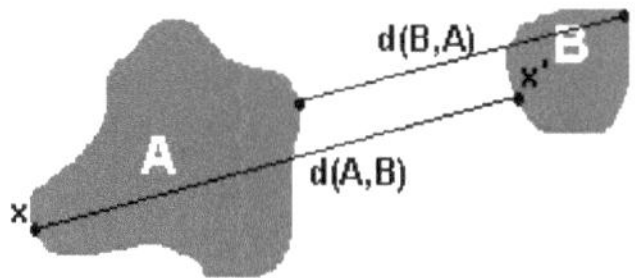

Then it is defined as follows:

The **distance from Hausdorf**:

Given A, B$\in \mathcal{H}(X)$, **h(A,B)=max{d(A,B), d(B,A)}** where (X,d) is a complete metric space.

(One can also denote $h(A,B)=d(A,B) \vee d(B,A)$.)

Proposition 1:

The Hausdorf distance is a metric on $\mathcal{H}(X)$.

Demonstration:

Let (X,d) be a complete metric space and A,B$\in \mathcal{H}(X)$.

    a) $h(A,B) = h(B,A), \forall$ A,B$\in \mathcal{H}(X)$?

$h(A,B) = \max\{ d(A,B), d(B,A)\} = \max\{ d(B,A), d(A,B)\} = h(B,A), \forall$ A,B$\in \mathcal{H}(X)$.

b) $0 < h(A,B) < \infty, \forall$ A,B$\in$ $\mathcal{H}(X)$, A$\neq$ B ?

Let A, B$\in$ $\mathcal{H}(X)$, such that A$\neq$ B.

As we have seen, $\exists x\in$ A,$\exists$ x'$\in$ B : d(A,B) = d(x,x') because A and B are non-empty compacts.

Since d is a metric on X, then $0\leq$ d(x,x') $<\infty$ and hence $0\leq$ d(A,B) $<\infty$,$\forall$ A,B$\in$ $\mathcal{H}(X)$. If A$\neq$ B, then $\exists x\in$ A: x$\notin$ B, whence d(A,B) $\neq 0$.

Thus, $0 < d(A,B) <\infty$ and also, $0 < d(A,B)\leq h(A,B) < \infty, \forall$ A,B$\in$ H(X), A$\neq$ B.

c)   h(A,A) = 0, $\forall$ A$\in$ $\mathcal{H}(X)$?

h(A,A) = max{d(A,A), d(A,A)} = max {d(a,A):a$\in$ A)} = =max{min{d(a,a'):a'$\in$ A}:a$\in$ A} = 0 because d(a,a)=0, $\forall$ a$\in$ A and A$\neq\varnothing$ and also d(x,y) $\geq 0, \forall$ x,y $\in$ A.

a)   h(A,B)$\leq$ h(A,C) + h(C,B),$\forall$ A, B, C$\in$ $\mathcal{H}(X)$?

Let us first see that d(A,B)$\leq$ d(A,C) + d(C,B),$\forall$ A, B, C$\in$ $\mathcal{H}(X)$.

Since A, B and C are compact and non-empty, we know that:

$\exists$ a$\in$ A, $\exists$ b$\in$ B : d(A,B) = d(a,b) ;

$\exists$ a'$\in$ A, $\exists$ c$\in$ C : d(A,C) = d(a',c) ;

$\exists$ b'$\in$ B,$\exists$ c''$\in$ C : d(C,B) = d(c',b').

We know that d(a',c) = d(A,C)$\geq$ d(x,C), $\forall$ x$\in$ A. In particular, d(A,C)$\geq$ d(a,C). (1)

Let c''$\in$ C: d(a,C) = d(a,c'') =min{d(a,z):z$\in$ C}.

Analogously:

$d(c',b') = d(C,B) \geq d(z,B), \forall z \in C.$

In particular, $d(C,B) \geq d(c'',B)$. (2)

Let $b'' \in B$: $d(c'',B) = d(c'',b'') = \min\{d(c'',y) : y \in B\}$.

So we have:

$$d(A,C) + d(C,B) \geq d(a,C) + d(c'',B) =$$

(1)
(2)

$$= d(a, c'') + d(c'',b'') \geq d(a,b'') \geq \min \{d(a,y) : y \in B\} = d(a,b) = d(A,B)$$

(X,d) metric space
. $a, c'', b'' \in X$ (because $A,B,C \subset X$)
. d checks triangular inequality

Therefore, $d(A,B) \leq d(A,C) + d(C,B), \forall\, A, B, C \in \mathcal{H}(X)$.

And, analogously, we can also prove that $d(B,A) \leq d(B,C) + d(C,A), \forall\, A, B, C \in \mathcal{H}(X)$.

So,

$$h(A,B) = \max \{ d(A,B), d(B,A)\} \leq \max \{ d(A,C) + d(C,B) , d(B,C) + d(C,A)\} \leq$$

$$\leq \max \{ d(A,C)\, d(C,A)\} + \max \{d(C,B) , d(B,C)\} = h(A,C) + h(C,B)$$

From any two parcels, one of each sum, I choose the larger one. Adding these two larger portions together gives me a sum that is greater than or equal to either of the previous two sums.

Therefore, $h(A,B) \leq h(A,C) + h(C,B), \forall\, A, B, C \in \mathcal{H}(X)$.

Therefore, the Hausdorf distance is a metric on $\mathcal{H}(X)$.

So,

Given a metric space (X, d) there exists a new metric space associated to it, with Hausdorf metric that can be represented by ($\mathcal{H}(X)$, h(d)).

Proposition 2:

If (X,d) is a complete metric space, then ($\mathcal{H}(X)$, h(d)) is a complete metric space.

Demonstration:

To demonstrate this proposition it will be necessary to introduce a definition and to demonstrate two lemmas beforehand.

Definition:

Let S be$\subset$ X and $\Gamma \geq 0$. Then, S $+\Gamma$ = {y$\in$ X : d(x,y) $\leq \Gamma$ for some x$\in$ S}. We call S $+\Gamma$ a dilation of S by a ball of radius $.\Gamma$

Motto 1:

Let A,B$\in$ $\mathcal{H}(X)$ where (X,d) is a metric space and letε > 0. Then, h(A,B) $\leq \varepsilon \Leftrightarrow$ A$\subset$ B+ $\varepsilon \wedge$ B$\subset$ A+ε .

Demonstration of Lemma 1:

h(A,B) = max { d(A,B), d(B,A}

( ) $\Rightarrow$

Suppose, without loss of generality, that h(A,B) = d(A,B) $\leq \varepsilon$.

Then, d(A,B) =max{d(a,B) : a A}$\in \leq \varepsilon$ $\Rightarrow$

$\Rightarrow$ d(a,B) $\leq \varepsilon$,$\forall$ a A $\in \Rightarrow$

$\Rightarrow \forall$ a$\in$ A,$\exists$ b$\in$ B : d(a,b) $\leq \varepsilon$ $\Rightarrow$

$\Rightarrow \forall$ to$\in$ A, to$\in$ B+ $\varepsilon \Rightarrow$

$\Rightarrow$ A$\subset$ B+ ε

If h(A,B) = d(A,B) $\leq \varepsilon$ then, by definition of h(A,B), comes d(B,A)$\leq$ d(A,B) $\leq \varepsilon$ and, analogously, one also has B$\subset$ A+ε .

(If we had assumed that h(A,B) = d(B,A), the whole demonstration of the lemma up to this point would be identical, exchanging A for B and vice versa.)

$(\ )\Leftarrow$

Suppose that $A \subset B + \varepsilon \wedge B \subset A + \varepsilon$.

$A \subset B + \varepsilon \Rightarrow \forall \ a \in A, a \in B + \varepsilon \Rightarrow$

$\Rightarrow \forall \quad a \in A, b \exists \in B : d(a,b) \leq \varepsilon \Rightarrow$

$\Rightarrow \quad \max \{d(a,B): a\ A\} \in \leq \varepsilon \Rightarrow$

$\Rightarrow \quad d(A,B) \leq \varepsilon$

Analogously, and if also $B \subset A + \varepsilon$, then $d(B,A) \leq \varepsilon$.

$d(A,B) \leq \varepsilon \wedge d(B,A) \leq \varepsilon \Rightarrow \max \{d(A,B), d(B,A)\} \leq \varepsilon \Rightarrow h(A,B) \leq \varepsilon$.

(End of the demonstration of Lemma. 1)

Lemma 2 (Extension Lemma):

Let (X,d) be an e.m. and let $\{A_n : n=1, 2, \ldots ,\infty \}$ a Cauchy succession of points in $(\mathcal{H}(X), h(d))$.

Let $\left\{n_{n_j}\right\}_{j=1}^{\infty}$, be an infinite succession of integers such that $0 < n < n_{12} < n < \ldots_3$

Supposing we have a Cauchy succession $\left\{x_{n_j} \in A_{n_j}\right\}_{j=1}^{\infty}$ on (X, d), then, there exists a Cauchy succession $\left\{\tilde{x}_n \in A_n\right\}_{n=1}^{\infty}$ such that $\tilde{x}_{n_j} = x_{n_j}, \forall j = 1,2,3,\ldots$.

Demonstration of Lemma 2:

The sequence $\left\{\tilde{x}_n \in A_n\right\}_{n=1}^{\infty}$ is constructed.

$\forall n \in \{1, 2, \ldots, n_1 \}$ we choose $\tilde{x}_n \in \left\{x \in A_n : d(x,x_{n_1}) = d(x_{n_1}, A_n)\right\}$ i.e., $\tilde{x}_n$ is the closest point (or one of the closest points) in A_n of x_{n_1} . This point actually exists because A_n is compact (since $A_n \in \mathcal{H}(X)$).

Analogously, $\forall j \in \{2,3,4...\}$ and $\forall n \in \{n_j +1,...,n_{j+1}\}$, we choose

$$\tilde{x}_n \in \left\{ x \in A_n : d(x, x_{n_j}) = d(x_{n_j}, A_n) \right\}$$

Let us now show that $\left\{ \tilde{x}_n \right\}_{n=1}^{\infty}$ has the desired properties, that is, that it is really an extension of $\left\{ x_{n_j} \right\}_{j=1}^{\infty}$ to $\left\{ A_n \right\}_{n=1}^{\infty}$.

Obviously, $\tilde{x}_{n_j} = x_{n_j}$ (because the closest point to x_{n_j} belonging to A_{n_j} is x_{n_j} itself) and also $\tilde{x}_n \in A_n$ by construction of the said succession.

To show that this is a Cauchy succession, let $\varepsilon > 0$ be a given value.

$\exists N_1 : n_k , n_j \geq N_1 \Rightarrow d(x_{n_k}, x_{n_j}) \leq \varepsilon /3$ (because $\left\{ x_{n_j} \right\}_{j=1}^{\infty}$ is a Cauchy succession by hypothesis).

Also $\exists N_2 : m, n \geq N_2 \Rightarrow d(A_m, A_n) \leq \varepsilon /3$ (because $\left\{ A_n \right\}_{n=1}^{\infty}$ is a Cauchy succession by hypothesis).

Let $N = \max \{N_1 , N_2 \}$.

Then, if $m, n \geq N$, $d\left(\tilde{x}_m, \tilde{x}_n \right) \leq d\left(\tilde{x}_m, \tilde{x}_{n_j} \right) + d\left(\tilde{x}_{n_j}, \tilde{x}_{n_k} \right) + d\left(\tilde{x}_{n_k}, \tilde{x}_n \right)$, in particular with $m \in \{n_{j-1} +1, n_{j-1} +2, ..., n_j \}$ and $n \in \{n_{k-1} +1, n_{k-1} +2, ..., n_k \}$.

If $h(A_m , A_{nj}) < \varepsilon /3$, then $\exists y \in A_m \cap (\{x_n \} + \varepsilon /3)$ such that $d(\tilde{x}_m, x_{n_j}) \leq \varepsilon /3$.

Analogously, $d(x_{n_k}, \tilde{x}_n) \leq \varepsilon /3$.

Hence, $d(\tilde{x}_m, \tilde{x}_n) \leq \varepsilon /3 + \varepsilon /3 + \varepsilon /3 = \varepsilon, \forall m, n \geq N$.

$\therefore \left\{ \tilde{x}_n \right\}_{n=1}^{\infty}$ is a Cauchy succession. (End of the proof of Lemma 2)

Let us now demonstrate the Proposition.

Let $\{A_n \}$ be a Cauchy succession on $H(X)$.

We will prove that $A = \lim_{n \to \infty} A_n \in H(X)$ and that $A = \{x \in X:$ there exists a Cauchy succession $\{x_n \in A_n \}$ that converges to $x\}$.

To this end we will break the demonstration into the following parts:

a) $A \neq \varnothing$

b) A is closed (and therefore will also be complete because $X \supset A$ and X is complete).

c) $\forall \varepsilon > 0, \exists N : n \forall \geq N, A \ A + \subset_n \varepsilon$

d) A is fully bounded and therefore, by b), will also be compact (i.e., $A \in H(X)$)

e) $\lim A_n = A$

Demonstration of a):

We will prove a) by showing the existence of a Cauchy succession $\{a \ A_i \in_i \}$ on X. For this we will find a succession of positive integers $N_1 < N < N_{23} < ... < N_n < ...$ such that $h(A_m, A_n) < \dfrac{1}{2^i}$, for m, n $> N$. i

Choose $x_{N_1} \in A_{N_1}$.

Then, as $h(A_{N_1}, A_{N_2}) \leq \dfrac{1}{2}$, we can find $x_{N_2} \in A_{N_2}$ such that $d\left(x_{N_1}, x_{N_2}\right) \leq \dfrac{1}{2}$

Suppose we select a finite sequence $x_{N_i} \in A_{N_i}$, i = 1, 2, ..., k, for which $d\left(x_{N_{i-1}}, x_{N_2}\right) \leq \dfrac{1}{2^{i-1}}$.

Then, as $h(A_{N_k}, A_{N_{k+1}}) < \dfrac{1}{2^k}$ and $x_{N_k} \in A_{N_k}$ we can find $x_{N_{k+1}} \in A_{N_{k+1}}$ such that $d\left(x_{N_k}, x_{N_{k+1}}\right) \leq \dfrac{1}{2^k}$.

For example, let $x_{N_{k+1}}$ be the point of $A_{N_{k+1}}$ closest to x_{N_k} .

By induction, we can find an infinite succession $\left\{x_{N_i} \in A_{N_i}\right\}$ such that $d\left(x_{N_i}, x_{N_{i+1}}\right) \leq \dfrac{1}{2^i}$.

To see that $\left\{x_{N_i}\right\}$ is a Cauchy succession on X, let $\varepsilon > 0$ and choose N_ε such that $\sum\limits_{i=N_\varepsilon}^{\infty} \dfrac{1}{2^i} < \varepsilon$.

Then, m$>$n $N\forall \geq_\varepsilon$ one has

$$d\left(x_{N_m}, x_{N_n}\right) \leq d\left(x_{N_m}, x_{N_{m+1}}\right) + d\left(x_{N_{m+1}}, x_{N_{m+2}}\right) + ... + d\left(x_{N_{n-1}}, x_{N_n}\right) <$$

$$< \sum\limits_{i=N_\varepsilon}^{\infty} d\left(x_{N_i}, x_{N_{i+1}}\right) = \sum\limits_{i=N_\varepsilon}^{\infty} \dfrac{1}{2^i} < \varepsilon$$

By the Extension Lemma (Lemma 2), there exists a convergent succession $\{a_i \in A_i\}$ for which $a_{N_i} = x_{N_i}$.

So, lim a_i exists and, by definition of A, that limit is in A.

Therefore, it is concluded that, $A \neq \varnothing$.

Demonstration of b):

To show that A is closed, suppose that $\{a_i \in A\}$ is a succession that converges to a point a.

Let us show that $\in$ A which will cause A to be closed.

$\forall i$, as $_i \in$ A, by definition of A, $\exists$ a succession $\{x_{i,n} \in A_n\}$ such that $\lim\limits_{n} x_{i,n} = a_i$.

There is also an increasing succession of positive numbers $\{N_i\}_{i=1}^{\infty}$ such that $d\left(a_{N_i}, a\right) \le \dfrac{1}{i}$.

Furthermore, there exists a succession of integers $\{m_i\}$ such that $d\left(x_{N_i, m_i}, a_{N_i}\right) \le \dfrac{1}{i}$

From here, $. d\left(x_{N_i, m_i}, a\right) \le d\left(x_{N_i, m_i}, a_{N_i}\right) + d\left(a_{N_i}, a\right) \le \dfrac{1}{i} + \dfrac{1}{i} = \dfrac{2}{i}$

Being $y_{mi} = x_{N_i, m_i}$ one has that $y_{mi} \in A_{m_i}$ and $\lim\limits_{i \to \infty} y_{mi} = a$.

By the Extension Lemma, $\{v_{mi}\}$ can be extended to a convergent succession $\{z_i \in A_i\}$ and then, in that case we will have a Cauchy succession that converges to a and, therefore, $a \in$ A (by definition of A) and we conclude that A is closed.

Demonstration of c):

$? \ \forall \varepsilon > 0, \exists \ N : n \ge N \Rightarrow A \subset A_n + \varepsilon \ ?$

Let $\varepsilon > 0$.

$\exists N : \forall \ m , n \ge N, \ h(A_m , A)_n \le \varepsilon$ because $\{A_n\}$ is Cauchy succession by hypothesis (see beginning of proposition demonstration)

Let n$\geq$ N. For m$\geq$ n, A $_m\subset$ A$_n$ +ε (by Lemma 1).

We want to show that A$\subset$ A$_n$ +ε .

For this, let be A.$\in$

(?$\Rightarrow$ a$\in$ A$_n$ +ε ?)

There is a succession $\{a_i \in A_i\}$ that converges to $\underline{a}$ (by definition of A, because a$\in$ A).

We can consider that N is also large enough so that for m$\geq$ N, d(a$_m$,a)$<\varepsilon$.

As A $_m\subset$ A$_n$ +ε , then $_m\in$ A$_n$ +ε .

Since A$_n$ is compact, it can also be shown that A$_n$ +ε is closed.

Since a $_m\in$ A$_n$ +ε , m$\forall\geq$ N, then $\underline{a}$ must also be in A $+_n\varepsilon$ because a$_m\to$ a and A$_n$ +ε is closed.

Demonstration of d):

? A is totally limited ?

Definition:

Let S be$\subset$ X. We say that S is **totally bounded** if$\forall\varepsilon$ >0, there exists a finite set of points $\{y , y_{12} , ..., y \}_n\subset$ S such that x$\forall\in$ X, y$\exists_i$: d(x,y)$<$.$_i\varepsilon$

The set $\{y , y_{12} , ..., y_n \}$ we call **network-** ε.

Suppose that A is not fully bounded.

Then, for givenε > 0, there is no finite net- ε.

Then we could find a succession $\{x_i\}_{i=1}^{\infty}$ in A such that d(x$_i$,x)$_j\geq\varepsilon$, i$\neq$ j. (3)

But, for c), there is n sufficiently large such that A$\subset$ A$_n$ +ε /3 which means that, for each x$_i$ there is y$_i$ corresponding in A$_n$ for which d(x$_i$,y)$_i\leq\varepsilon$ /3. (4)

Since A$_n$ is compact, $\{y_i\}_{i=1}^{\infty}$ has a convergent subsucession $\{y_{n_i}\}$ on A$_n$.

So we can find points of succession $\{y_{n_i}\}$ as close as we like.

In particular, we can find two points y_{n_i} and y_{n_j} such that $d(y_{n_i}, y_{n_j}) < \dfrac{\varepsilon}{3}$. (5)

But in that case,

$$d(x_{n_i}, x_{n_j}) \leq d(x_{n_i}, y_{n_i}) + d(y_{n_i}, y_{n_j}) + d(y_{n_j}, x_{n_j}) \leq \frac{\varepsilon}{3} + \frac{\varepsilon}{3} + \frac{\varepsilon}{3} = \varepsilon$$

(4) (5) (4)

which contradicts (3).

Then A is totally bounded and, since by b) it is also closed, then A is compact, that is, $A \in$ H(X) (because it is also non-empty, by a)).

Demonstration of e):

$A \in$ H(X)

By c), $\forall \varepsilon > 0 \, \exists$ N : $n \geq N \Rightarrow A \subset A_n + \varepsilon$.

If we prove that also $\forall \varepsilon > 0, \exists$ N : $n \geq N \Rightarrow A_n \subset A + \varepsilon$, then, by Lemma 1, we obtain that $\forall \varepsilon > 0, \exists$ N : $n \geq N \Rightarrow h(A_n , A) < \varepsilon$, whence we get $\lim A_n = A$.

Let $\varepsilon > 0$.

Let us find N such that $\forall$ m,n$\geq$ N, h(A$_m$, A)$_n \leq \varepsilon$ /2.

Then, by Lemma 1, $\forall$ m,n$\geq$ N, A$_m \subset$ A $+_n \varepsilon$ /2. (6)

Let n be$\geq$ N. We want to show that $A_n \subset A + \varepsilon$.

Let y $A \in _n$.

There exists an increasing succession $\{N_i\}$ of integers such that N < N < N$_{12}$ < ... < N$_k$ < ... and for m,n$\geq$ N$_j$, A $_m \subset$ A$_n + \varepsilon$ / 2 $.^{j+1}$

Note that A $_n \subset A_{N_1} + \dfrac{\varepsilon}{2}$ (because n> N$_1$ and by (6)).

As y $A \in _n$, $. \exists x_{N_1} \in A_{N_1} : d(y, x_{n_1}) \leq \dfrac{\varepsilon}{2}$

As $, . x_{N_1} \in A_{N_1} \exists x_{N_2} \in A_{N_2} : d(x_{N_1}, x_{n_2}) \leq \dfrac{\varepsilon}{2^2}$

Similarly and by mathematical induction we can find a succession $x_{N_1}, x_{N_2}, x_{N_3}, ...$ such that $x_{N_j} \in A_{N_j}$ e $d(x_{N_j}, x_{N_{j+1}}) \leq \dfrac{\varepsilon}{2^{j+1}}$.

Using the triangular inequality, a given number of times, we can show that

61

$$d(y, x_{n_j}) \le \sum_{j=1}^{\infty} \frac{\varepsilon}{2^{j+1}} = \varepsilon \cdot \frac{1}{2^2} \times \frac{1}{1 - \frac{1}{2}} = \frac{\varepsilon}{2}, \forall j$$

and also $\left\{ x_{N_j} \right\}$ is a Cauchy succession.

The way n was chosen, each $A_{N_j} \subset A_n + \varepsilon /2$.

$\left\{ x_{N_j} \right\}$ converges to a point x and since $A_n + \varepsilon /2$ is closed, then also x $\in A_n + \varepsilon /2$ hence $d(x, x_{N_j}) \le \frac{\varepsilon}{2}$ for N sufficiently large.

In addition, $d(y, x) \le d(y, x_{N_j}) + d(x_{N_j}, x) \le \frac{\varepsilon}{2} + \frac{\varepsilon}{2} = \varepsilon$. (7)

By the Extension Lemma, there exists a Cauchy succession $\left\{ x_n \in A_n \right\}$ that converges to x.

Therefore, $x \in A$ (by definition of A).

And, from (7) we can conclude that y $\in A + \varepsilon$ because $\exists x \in A : d(y,x) \le \varepsilon$.

Therefore, $A_n \subset A + \varepsilon$, for n $\ge$ N.

And so, since $\exists$ N : $\forall$ n>N $A_n \subset A + \varepsilon \wedge A \subset A_n + \varepsilon$, then h(A_n ,A) $\le \varepsilon$, $\forall$ n>N, that is $\lim A_n = A$.

Therefore, (H(X) , h(d)) is a complete metric space. $\blacklozenge$

H(X) will then be the space where the fractals "live".

A fractal will be, for the time being, a compact non-empty subset of a complete metric space.

Sierpinsky Triangle - This triangle is constructed in the following way: Starting with a triangle identical to the one in the previous figure, mark the midpoints of each of its sides which are joined by segments dividing it into four new smaller triangles similar to the initial one. Remove the triangle in the middle and do the same process for each of the remaining triangles. And so on... The picture shows the result after performing this operation three times, but is intended to represent the set that would be obtained if we performed this task infinitely many times. (Only in our mind can we conceive of it. We can't represent it graphically, because of time, space and computer screen resolution). →

We will continue the construction of the concept of a fractal as a fixed point of a system of iterated functions, but the idea of a fractal is even today easier to understand through various figures and contexts than by a formal definition.

What is a fixed point of a transformation?

Definition:

Let $f: X \rightarrow X$ be an application of an e.m. on itself.

$x_f \in X$ is a fixed point of f if $f(x_f) = x_f$

To arrive at fractals we are interested, among the transformations of X, in those that are **contractions**.

Definition:

A transformation $f: X \rightarrow X$ on an e.m. (X,d) is said to be a contraction if there exists a constant s, with $0 \leq s < 1$, such that :

$$d(f(x), f(y)) \leq s \cdot d(x,y) , \forall\ x,y \in X.$$

We call this the shrinkage factor.

Examples:

1. X = C [0,1] (Set of continuous functions f: [0,1]$\rightarrow$ IR)

A "point" of X is a function.

Metric defined in this space: $d(f,g) = \max\left\{ |f(x)-g(x)| : x \in [0,1] \right\}$ (it is well defined because f and g are continuous on [0,1])

Example of contraction in

C [0,1]:

w: C [0,1]$\rightarrow$ C [0,1]

f$\rightarrow$ w(f) = ½ f +   1

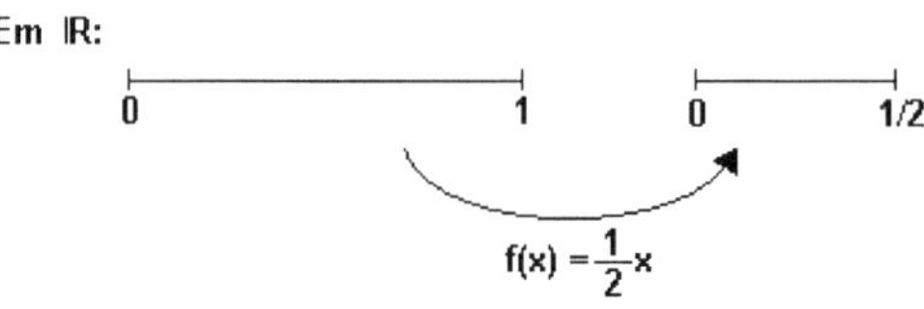

whereby

(½ f + 1)(x) = ½ f(x) + 1, x$\forall\in$ [0,1]

Contraction factor : ½ **Fixed point :** p(x) = 2 , x$\forall\in$ [0,1]

2.

Shrinkage factor : ½   **Fixed point :** 0

3. w: T $_s\to$ T$_s$ (T$_s$ - Sierpinski's Triangle)

x$\to$ w(x) = Ax + t

with

$$A = \begin{bmatrix} \dfrac{1}{2}\cos 120^o & -\dfrac{1}{2}sen120^o \\[2ex] \dfrac{1}{2}sen120^o & \dfrac{1}{2}\cos 120^o \end{bmatrix}$$

$$e\; t = \begin{bmatrix} \dfrac{1}{2} \\[2ex] 0 \end{bmatrix}$$

Shrinkage factor : ½

Fixed point: $\left(\dfrac{5}{14}; \dfrac{\sqrt{3}}{14} \right)$

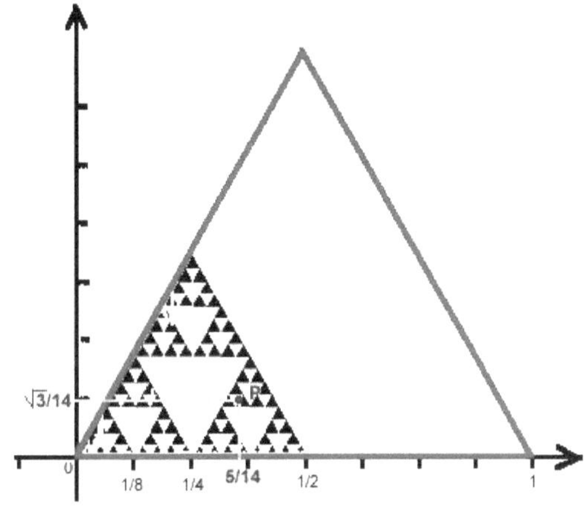

This application could more simply be defined from IR^2 on IR^2 , but we can also consider that the Sierpinski triangle is a subspace of IR^2 , that is, it is itself a metric space. The application defined in this way will be useful to us for an example later on.

The fractal nature of space, that is, its self-similarity to infinity, allows us to see that space is applied within itself.

Applying this transformation, the space is reduced to half its size, rotated 120° to the left around the origin and translocated ½ unit horizontally to the right.

Proposition 3:

If f,g: X$\to$ X are contractions of contraction factors s and t respectively, then fog is a contraction of contraction factor st.

Demonstration:

f is contraction factor $s\Rightarrow$ d(f(x),f(y))$\leq s$. d(x,y),$\forall$ x,y$\in$ X (8)

g is contraction factor $t\Rightarrow$ d(g(x),g(y))$\leq t$. d(x,y),$\forall$ x,y$\in$ X (9)

d(fog(x), fog(y)) = d(f(g(x)), f(g(y)))$\leq s$. d(g(x), g(y))$\leq st$. d(x,y),$\forall$ x,y$\in$ X

$$\qquad\qquad\qquad\qquad\qquad\qquad\quad \boxed{(8)} \qquad\qquad\qquad \boxed{(9)}$$

$0\leq s < 1 \wedge 0\leq t < 1 \Rightarrow 0\leq st < 1$. Therefore, st is contraction factor of f **O** g.

6.3 - DEFINITION OF ITERATED FUNCTION SYSTEM

Let us now define what we need to build the concept of Fractal.

Definition:

Let f: $X \to X$ be a transformation on a metric space.

Successive iterations of f are transformations f^{on} : $X \to X$ defined by:

$f^{o0}(x) = x$

$f^{o1}(x) = f(x)$

$f^{o(n+1)}(x) = (f \circ f^{on})(x) = f(f^{on}(x))$, n = 0, 1, 2, 3, ...

If we now think about successive iterations of a contraction, we can prove that:

Proposition 4:

If f: $X \to X$ is a contraction of contraction factor s, then f^{on} : $X \to X$ is a contraction of contraction factor s^n .

Demonstration:

We simply apply the previous proposition n-1 times.

Proposition 5:

If f: $X \to X$ is a contraction defined on a complete metric space. (X,d), then f has exactly one fixed point $x_f \in X$ and, furthermore, for any point $x \in X$, the sequence $\{f^{on}(x): n = 0, 1, 2,...\}$ converges to x_f, that is, . $\lim_{n \to \infty} f^{on}(x) = x_f, \forall x \in X$

Demonstration:

I will divide this demonstration into three parts:

    a) $\{f^{on}(x)\}_{n=0}^{\infty}$ is a Cauchy succession, x $X \forall \in$

    b) $\{f^{on}(x)\}_{n=0}^{\infty}$ converges to the fixed point of f.

    c)   f has exactly one fixed point.

Demonstration of a):

Let x be $\in$ X, let s be the contraction factor of f $(0 \leq s < 1)$

$\{f^{on}(x)\}_{n=0}^{\infty}$ is a Cauchy succession, x $X \forall \in \Leftrightarrow \Leftrightarrow$ $\forall \varepsilon > 0, \exists$ N : $\forall$ m,n > N $\Rightarrow$ d(f^{on} (x) , $f^{om}(x)) < \varepsilon$

Now, fixed a point $x \in X$,

$$d(f^{on}(x), f^{om}(x)) \leq s\min\{m,n\} \cdot d(x, f^{o|m-n|}(x)), \forall\, m,n = 0, 1, 2, \ldots \quad (10)$$

In particular, for $k = 0, 1, 2, \ldots$, we have:

```
┌────────────────────────┐
│ Triangular inequality  │
└────────────────────────┘
```

$$d(x, f^{ok}(x)) \leq d(x, f(x)) + d(f(x), f^{o2}(x)) + d(f^{o2}(x), f^{o3}(x)) + \ldots + d(f^{o(k-1)(x)}, f(x))^{ok}$$

```
┌──────────────────┐
│ f is contraction │
└──────────────────┘
```

$$= d(x, f(x)) + s \cdot d(x, f(x)) + s \cdot d(f(x), f^{o2}(x)) + \ldots + s \cdot d(f^{o(k-2)}(x), f(x))^{o(k-1)}$$

```
┌──────────────────┐
│ ≤                │
│ Geometric        │
│ series of ratio s│
│ (|s|<1)          │
└──────────────────┘
```

$$= (1 + s + s^2 + \ldots + s^{k-1}) \cdot d(x, f(x))$$

$$= (1 - s)^{-1} \cdot d(x, f(x))$$

Turning now to (10) we have:

$$\forall\, m,n = 0, 1, 2, \ldots, \quad d(f^{on}(x), f^{om}(x)) \leq s\min\{m,n\} \cdot (1 - s)^{-1} \cdot d(x, f(x))$$

where $d(x, f(x)) \in \mathbb{R}_0^+$.

Come on,

$$s\min\{m,n\} \cdot (1 - s)^{-1} \cdot d(x, f(x)) < \varepsilon \Leftrightarrow$$

$$\Leftrightarrow s^{\min\{m,n\}} < \frac{\varepsilon(1-s)}{d(x, f(x))} \Leftrightarrow$$

$$\Leftrightarrow s^{\min\{m,n\}} < s^{\log_s\left(\frac{\varepsilon(1-s)}{d(x,f(x))}\right)} \Leftrightarrow$$

$$\Leftrightarrow \min\{n,m\} < \log_s\left(\frac{\varepsilon(1-s)}{d(x, f(x))}\right)$$

Then, given $x \in X$ fixed and given $\varepsilon > 0$, it is enough to choose N the smallest integer such that $N > \log_s\left(\dfrac{\varepsilon(1-s)}{d(x, f(x))}\right)$.

Therefore, $\{f^{on}(x)\}_{n=0}^{\infty}$ is a Cauchy succession, $\forall\, x \in X$.

Demonstration of b):

Since X is, by hypothesis, a complete metric space, then $\{f^{on}(x)\}^{\infty}_{n=0}$ has boundary on X - let's call it x_f.

$$\lim_{n\to\infty} f^{on}(x) = x_f \in X$$

x_f is a fixed point of f?

$$f(x_f) = f\left(\lim_{n\to\infty} f^{on}(x)\right) = \lim_{n\to\infty} f\left(f^{on}(x)\right) = \lim_{n\to\infty} f^{o(n+1)}(x) = x_f$$

Is a subsucession of $f^{on}(x)$

Therefore, x_f is a fixed point of f.

Demonstration of c):

x_f is the only fixed point of f?

Suppose that f has two fixed points x_f and y_f.

Then, $f(x_f) = x_f$ and $f(y_f) = y_f$ which makes $d(f(x_f), f(y_f)) = d(x_f, y_f)$.

On the other hand, since f is a contraction of contraction factor s, then:

$d(f(x_f), f(y_f)) \le s.d(x_f, y_f)$.

Whence, $d(x_f, y_f) \le s.d(x_f, y_f) \Rightarrow$

$\Rightarrow (1-s).d(x_f, y_f) \le 0 \Leftrightarrow$

$(1-s) > 0$ because $0 \le s < 1$
$d(x_f, y_f) \ge 0$ because d is metric

$\Leftrightarrow d(x_f, y_f) = 0 \Leftrightarrow$

$\Leftrightarrow x_f = y_f$

d is metric

Therefore, x_f is the only fixed point of f.

Therefore, we conclude that, $\forall x \in X$, $\{f^{on}(x)\}$ is a convergent succession for x_f which is the only fixed point of f.

Example:

Using the example already presented in space C [0,1]:

w: C [0,1] $\to$ C [0,1]

f $\to$ w(f) = ½ f + 1

where $(½ f + 1)(x) = ½ f(x) + 1$, $\forall x \in [0,1]$

Contraction factor : ½   **Fixed point :** $p(x) = 2$, $\forall x \in [0,1]$

We know that any succession of successive iterations of any function f from C[0, 1] tends to the function p(x) = 2 , x∀∈ [0, 1].

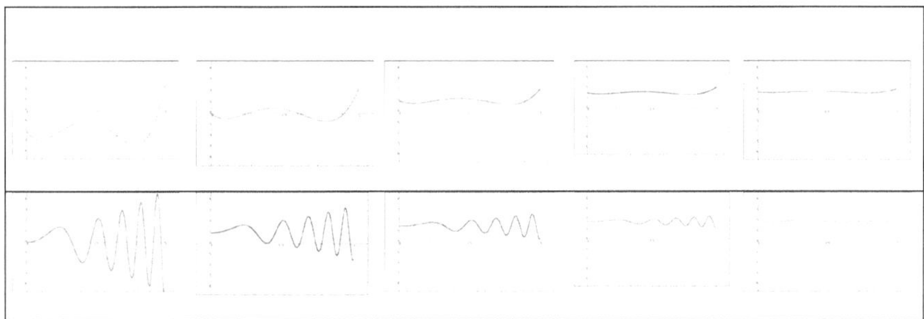

Two examples presenting the first four iterations by w of two functions of the space C[0,1], iterations that approach the fixed point of w.

If we now think of **X as a point of** $\mathcal{H}(X)$, as ($\mathcal{H}(X)$, h(d)) is a complete metric space, if $\tilde{w}$: $\mathcal{H}(X)\to \mathcal{H}(X)$ is a contraction on that metric space. Then, analogously, we can state that, $\left\{ \tilde{w}^{\,\overset{on}{\sim}} (X) \right\}_{n=0}^{\infty}$ is a Cauchy sequence in ($\mathcal{H}(X)$, h(d)) and that $\lim_{n\to\infty} \tilde{w}^{\,\overset{on}{\sim}} (X) = \left\{ x_{w} \right\}$

where x_{w} is the fixed point of $\tilde{w}$.

We only need to define this contraction in ($\mathcal{H}(X)$, h(d)).

Proposition 6:

Let w: X→ X be a contraction with contraction factor s on a metric space X, and let $\tilde{w}$: $\mathcal{H}(X)\to \mathcal{H}(X)$ such that , B $w(B) = \{w(x): x \in B\} \forall \in \mathcal{H}(X)$.

So:

    a) w is continuous,

    b) $\tilde{w}$ applies $\mathcal{H}(X)$ to itself,

    c) $\tilde{w}$ is a contraction in ($\mathcal{H}(X)$, h(d)) with contraction factor s.

Demonstrations:

Demonstration of a):

w is continuous $\Leftrightarrow \forall \varepsilon > 0, \exists \delta > 0 : d(x,y) < \delta \Rightarrow d(w(x), w(y)) < \varepsilon$

Let be $\varepsilon > 0$

$d(x,y) < \delta \Rightarrow d(w(x), w(y)) \leq s \cdot d(x,y) < s \cdot \delta$

If $\delta = \varepsilon / s$, then $\forall \varepsilon > 0, \exists \delta = \varepsilon / s > 0 : d(x,y) < \delta \Rightarrow d(w(x), w(y)) < \varepsilon$

Therefore, if w is a contraction, then w is a continuous application on X.

Demonstration of b):

We know that $\tilde{w}$ applies $\mathcal{H}(X)$ on itself if we prove that $\tilde{w}(S) \in H(X)$, $S \forall \in \mathcal{H}(X)$.

Let S be $\in \mathcal{H}(X)$.

Then, S is a compact non-empty subset of X.

$\tilde{w}(S) = \{w(x) : x \in S\}$ is non-empty since $S \neq \varnothing$.

? $\tilde{w}(S)$ is compact?

$\tilde{w}(S)$ is compact if its entire succession has a convergent subsection at $\tilde{w}(S)$.

Let $\{x_n\}_{n=1}^{\infty}$ be a succession in S.

Since S is compact, then $\{x_n\}$ has a subsection that is convergent on S - say $\{x_{N_n}\}$.

Let $\{y_N = w(x_N)\}_{n=1}^{\infty}$ be a succession on $\tilde{w}(S)$.

Since w is continuous and $\{x_{N_n}\}$ is convergent on S, then $\{y_{N_n} = w(x_{N_n})\}_{n=1}^{\infty}$ is also convergent on $\tilde{w}(S)$.

Therefore, $\tilde{w}(S)$ is also compact and $\tilde{w}$ applies $\mathcal{H}(X)$ on itself.

Demonstration of c):

? $\tilde{w}$ is contraction in $(\mathcal{H}(X), h(d))$ with contraction factor s?

Let A, B $\in \mathcal{H}(X)$.

$h(\tilde{w}(A), \tilde{w}(B)) = \max \{d(\tilde{w}(A), \tilde{w}(B)) ; d(\tilde{w}(B), \tilde{w}(A))\}$

$d(\tilde{w}(A), \tilde{w}(B)) = \max \{d(w(x), \tilde{w}(B)) : x \in A\} =$

$$= \max \{\min \{d(w(x),w(y)) : y \in B\} : x \in A\} \le$$

$$\le \max \{\min \{s.d(x,y) : y \in B\} : x \in A\} =$$

$$= \max \{s. \min \{d(x,y) : y \in B\} : x \in A\} =$$

$$= \max \{s. d(x,B) : x \in A\} =$$

$$= s. \max \{d(x,B) : x \in A\} =$$

$$= s. d(A,B)$$

Analogously, $d(\tilde{w}(B), \tilde{w}(A)) \le s. d(B,A)$

Then, $h(\tilde{w}(A), \tilde{w}(B)) \le \max \{ s. d(A,B) ; s. d(B,A)\} =$

$$= s. \max \{d(A,B) ; d(B,A)\} =$$

$$= s. h(A,B)$$

Therefore, we conclude that, $\tilde{w}$ is a contraction with contraction factor s.

Examples:

1. Using the example already given in IR, $f(x)=1/2\ x$, whose fixed point is $x = 0$, the previous result tells us that $\tilde{f}^{on}([0,1]) = \{0\}$.

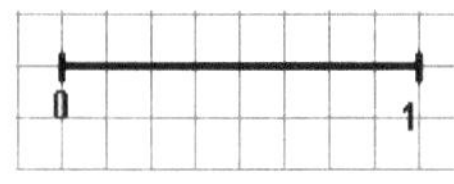
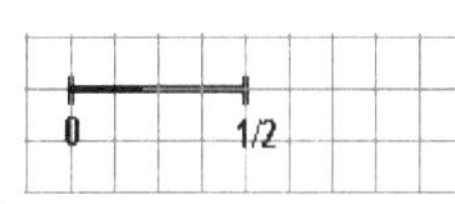
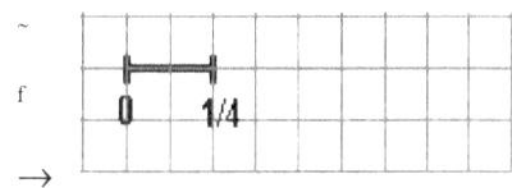

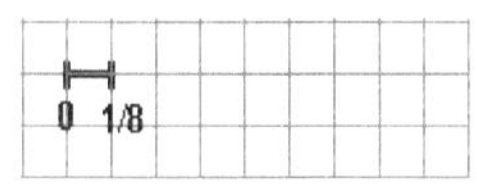
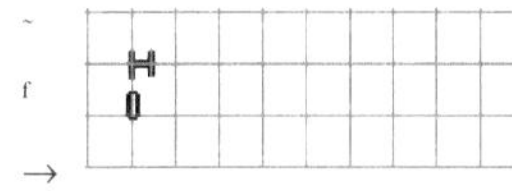
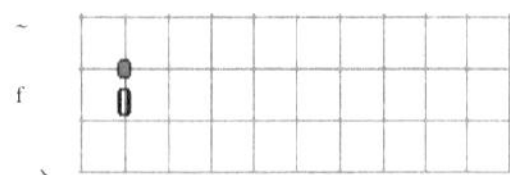

2. Taking now the other example defined in the Sierpinski Triangle, whose fixed point is $\left(\dfrac{5}{14}, \dfrac{\sqrt{3}}{14}\right)$, and using the previous result, we know that $w(x)$ defined on T_s induces an application $\tilde{w}$ on $\mathcal{H}(T_s)$ such that the succession of successive iterations from any point of this e.m., namely, to T itself$_s$, converges to $\left\{\left(\dfrac{5}{14}, \dfrac{\sqrt{3}}{14}\right)\right\}$.

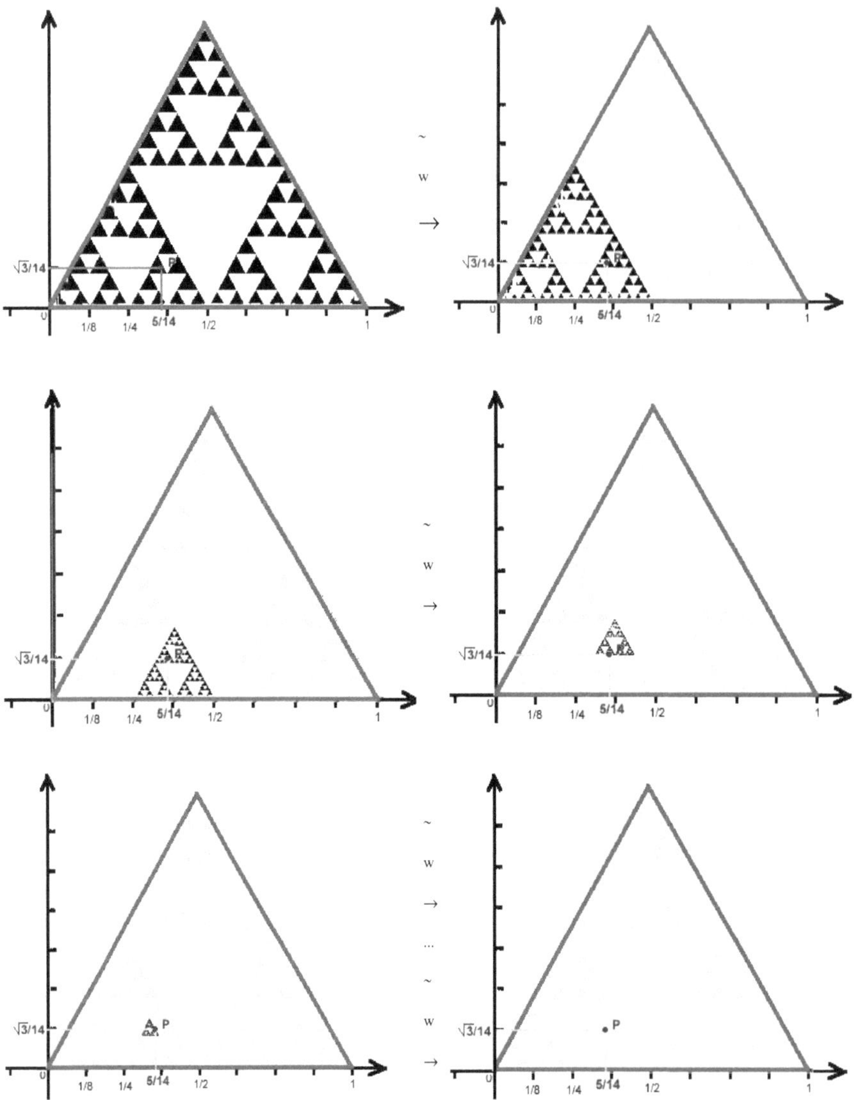

This result, although important, is of little practical interest, at least in graphical terms, since in the end, that is, in the limit of infinitely many successive iterations of any point of H(X), we obtain one and only one point of X.

But what if instead of using only one contraction at a time, we operate with several contractions simultaneously?

Proposition 7:

Let (X,d) be a metric space

Let $\{\tilde{w}_n : n= 1, 2, ..., N\}$ be a set of contractions in $(\mathcal{H}(X), h(d))$ each with its respective contraction factor $s_n \forall\, n=1, 2, ..., N$.

Let $W:\mathcal{H}(X)\to \mathcal{H}(X)$ be such that

$$W(B) = \tilde{w}_1(B) \cup \tilde{w}_2(B) \cup ... \cup \tilde{w}_N(B) = \, B\bigcup_{n=1}^{N} \tilde{w}_n(B)\, \forall\in\, \mathcal{H}(X).$$

Then W is a contraction in $\mathcal{H}(X)$ with contraction factor

$s = \max\{s_n : n= 1, 2, ..., N\}$

Demonstration:

Let us use the method of mathematical induction.

Hypothesis: $\{\tilde{w}_n : n= 1, 2, ..., N\}$ is a set of contractions in $(\mathcal{H}(X), h(d))$ each with its respective contraction factor $s_n \forall\, n=1, 2, ..., N$.

Thesis: $P(N)$: $W:\mathcal{H}(X)\to \mathcal{H}(X)$ such that

$$W(B) = \tilde{w}_1(B) \cup \tilde{w}_2(B) \cup ... \cup \tilde{w}_N(B) = \, B\bigcup_{n=1}^{N} \tilde{w}_n(B)\, \forall\in\, \mathcal{H}(X) \text{ is a contraction on}$$

$\mathcal{H}(X)$ with contraction factor $s = \max\{s_n : n= 1, 2, ..., N\}$.

For $N-2$:

? $Q(2)$: $W_2 :\mathcal{H}(X)\to \mathcal{H}(X)$ such that $W_2(B) = \tilde{w}_1(B) \cup \tilde{w}_2(B) = \, B\bigcup_{n=1}^{2} \tilde{w}_n(B)\, \forall\in$

$\mathcal{H}(X)$ is a contraction on $\mathcal{H}(X)$ with contraction factor $s = \max\{s_n : n= 1, 2\}$?

Let B, C be$\in \mathcal{H}(X)$.

$$h(W_2(B), W_2(C)) = h(\,(B)\tilde{w}_1\cup\tilde{w}_2(B), \tilde{w}_1(C)\cup\tilde{w}_2(C))$$

For the purposes of writing simplification, be:

$$B_1 = \tilde{w}_1(B)\,;\; B_2 = \tilde{w}_2(B)\,;\; C =_1 \tilde{w}_1(C)\,;\; C_2 = \tilde{w}_2(C).$$

Then, $h(B\ B_{1\cup 2}, C\ C_{1\cup 2}) = \max\{\ d(B\ B, C_{1\cup 21\cup 2}),\ Cd(C\ C_{1\cup 2}, B\ B_{1\cup 2})\}.$

Come on,

$d(B\ B_{1\cup 2}, C\ C_{1\cup 2}) = \max\{d(x, C\ C_{1\cup 2}): x\in B\ B_{1\cup 2}\} =$

$= \max\ \{\ \max\{d(x, C\ C_{1\cup 2}): x\in B_1\ \}\ ;\ \max\{d(x, C\ C_{1\cup 2}): x\in B$
$\}\}=_2$

$= \max\ \{\ d(B_1, C\ C_{1\cup 2}), d(B_2, C\ C_{1\cup 2})\ \} =$

$= \max\ \{\ \min\ \{d(B_1, C_1), d(B_1, C_2)\}, \min\ \{d(B_2, C_1), d(B_2, C_2)\}\}=$

$\leq \max\ \{\ d(B_1, C_1), d(B_2, C_2)\ \}$

Analogously, $d(C\ C_{1\cup 2}, B\ B_{1\cup 2}\leq\ )\max\ \{\ d(C_1, B_1), d(C_2, B_2)\ \}$

So,

$h(W_2\ (B), W_2\ (C)) = h(\ (B)\ \tilde{w}_1\cup \tilde{w}_2\ (B), \tilde{w}_1\ (C) \cup \tilde{w}_2\ (C)) =$

$= h(B\ B_{1\cup 2}, C\ C\ )_{1\cup 2}\leq \max\ \{\ d(B, C_{11}), d(B, C_{22}), d(C, B_{11}), d(C, B_{22})\} =$

$= \max\ \{\ \max\ \{d(B_1, C_1), d(C_1, B_1)\}, \max\{d(B_2, C_2), d(C_2, B_2)\}\} = =$

$= \max\ \{h(B_1, C_1), h(B_2, C_2)\} = =$

$= \max\ \{h(\tilde{w}_1\ (B), \tilde{w}_1\ (C)), h(\tilde{w}_2\ (B), \tilde{w}_2\ (C))\} \leq$

$\leq \max\ \{s_1\ .\ h(B,C), s_2\ .\ h(B,C)\} = s\ .\ h(B,C)$

$\boxed{s = \max\ \{s_1, s\ \}_2}$

$\boxed{\begin{array}{l}\tilde{w}_1 \text{ is contraction with factor } s_1 \\ \tilde{w}_2 \text{ is contraction with factor } s_2\end{array}}$

Therefore, W_2 is a contraction of contraction factor s and we demonstrate that P(2) is true.

Suppose now that the proposition P(k) is true, that is, that when $N = k$ the following is true:

Hypothesis: P(k): $W:\mathcal{H}(X)\to \mathcal{H}(X)$ such that

$W_k\,(B) = \tilde{w}_1\,(B) \cup \tilde{w}_2\,(B) \cup \,...\cup\, \tilde{w}_k\,(B) = ,\ B \bigcup\limits_{n=1}^{k} \tilde{w}_n(B)\ \forall \in\ \mathcal{H}(X)$ is a contraction on $\mathcal{H}(X)$ with contraction factor $s_k = \max\,\{s_n : n = 1, 2, ..., k\}.$

Let us now prove that the proposition will also be true for $N = k+1$.

Thesis: $P(k+1)$: $W : \mathcal{H}(X) \to \mathcal{H}(X)$ such that

$W_{k+1}\,(B) = \tilde{w}_1\,(B) \cup \tilde{w}_2\,(B) \cup \,...\cup\, \tilde{w}_k\,(B) \cup \tilde{w}_{k+1}\,(B) = ,\ B \bigcup\limits_{n=1}^{k+1} \tilde{w}_n(B)\ \forall \in\ \mathcal{H}(X)$ is a

contraction on $\mathcal{H}(X)$ with contraction factor $s = \max\,\{s_n : n = 1, 2, ..., k, k+1\}.$

Let then be , , ... , ... , $\tilde{w}_1\,\tilde{w}_2\,\tilde{w}_k\,\tilde{w}_{k+1} :\ ,\mathcal{H}(X) \to \mathcal{H}(X)$ the contractions with contraction factors s , s_{12} , ... , s , s_{kk+1} respectively that define $W_{.k+1}$.

$\forall B, C \in\ \mathcal{H}(X),$

$h\,(W_{k+1}\,(B), W_{k+1}\,(C)) =$

$= h\,(\tilde{w}_1\,(B) \cup \tilde{w}_2\,(B) \cup \,...\cup \tilde{w}_k\,(B) \cup \tilde{w}_{k+1}\,(B)\,,\, \tilde{w}_1\,(C) \cup \tilde{w}_2\,(C) \cup \,...\cup \tilde{w}_k\,(C) \cup \tilde{w}_{k+1}\,(C)\,) =$

$= h\,(W_k\,(B) \cup \tilde{w}_{k+1}\,(B)\,,\, W_k\,(C) \cup \tilde{w}_{k+1}\,(C)\,) \leq$ $\boxed{P(2)}$

$\leq \max\,\{\,h\,(W_k\,(B)\,,\,W_k\,(C))\,,\,h(\tilde{w}_{k+1}\,(B)\,,\,\tilde{w}_{k+1}\,(C))\,\} \leq$ $\boxed{P(k)}$

$\leq \max\,\{s_k\,h(B,C)\,,\,s_{k+1}\,h(B,C)\} =$ $\tilde{w}_{k+1}$ is contraction of factor s_{k+1}

$S = \max\,\{s_1\,,\,s_2\,,\,...\,,\,s_k\,,\,s\,\}_{k+1}$

$= s\,.\,h(B,C)$

$\therefore P(2) \wedge P(k) \Rightarrow P(k+1)$

$\therefore$ $P(N)$ is true, $N \forall \in$ IN, that is, W is contraction in $\mathcal{H}(X)$ with contraction factor s.

We have just defined what is required to obtain an **Iterated Function System (IFS)** consisting of a metric space (X,d) together with a finite set of contractions $w_n : X \to X$ with the corresponding contraction factors s_n , $n=1,2,...,N$.

We denote this IFS by {**X; w$_n$, n=1,2,...,N**} and its **contraction factor** is **s = max {s$_n$: n= 1, 2, ..., N}**.

Since W (defined as in the previous proposition, is a contraction in $\mathcal{H}(X)$, we already know that $\left\{W^{on}(B)\right\}_{n=1}^{\infty}$ is a Cauchy succession in $\mathcal{H}(X)$, which converges to the fixed point of W, $B \forall \in \mathcal{H}(X)$. We call this fixed point a **fractal**.

To recap:

Let (X,d) be a complete metric space.

Let {X; w$_n$, n=1,2,...,N} be an IFS on X, consisting of N contractions, each with contraction factor s$_n$ respectively - the contraction factor of the IFS is s = max {s$_n$: n= 1, 2, ..., N}.

Each contraction w$_n$: X→ X induces a contraction $\tilde{w}_n$: H(X)→ H(X) with equal contraction factor such that , B $\tilde{w}(B) = \{w(x) : x \in B\} \forall \in$ H(X).

Together, these form a new contraction W:H(X)→ H(X) with contraction factor s = max {s$_n$: n= 1, 2, ..., N} defined by

$$W(B) = \tilde{w}_1(B) \cup \tilde{w}_2(B) \cup ... \cup \tilde{w}_N(B) = , B \bigcup_{n=1}^{N} \tilde{w}_n(B) \forall \in H(X).$$

Since W is a contraction on (H(X), h(d)) which is a complete metric space, then we know that this will have a single fixed point A∈ H(X) and that the succession $\left\{W^{on}(B)\right\}_{n=1}^{\infty}$ converges to A, $B\forall \in$ H(X). A is also said to be the fixed point of the IFS.

$$A = W(A) = \bigcup_{n=1}^{N} w_n(A) \text{ and } A = B \lim_{n \to \infty} W^{on}(B), \forall \in H(X).$$

We call this subset A of X **FRACTAL**.

Like all mathematical entities, fractals only exist in a pure state in our mind, as it is impossible in time and space to concretise them graphically, since we cannot determine an infinite number of iterations of a function.

The computer is a very valuable tool in the study and construction of fractals, since it allows us to quickly perform a huge amount of calculations and to represent them graphically with great precision, also giving us the possibility to easily observe in the figure the implications of changing some parameters in the expressions of the functions that make up the IFS in question. Even so, and no matter how powerful the machine is, its capacity for calculation and graphical representation is always limited and we must take that into account, imagining what the machine cannot perform.

6.4 - PRIMITIVE EXAMPLES OF FRACTALS

6.4.1 - SIERPINSKI TRIANGLE

I use this example again, this time introducing the IFS that produces it.

IFS: $\left\{ IR^2;\, w_1,\, w_2, w_3 \right\}$ where :

$$w_1(x,y) = \left(\frac{1}{2}x, \frac{1}{2}y \right) \quad w_2(x,y) = \left(\frac{1}{2}x, \frac{1}{2}y \right) + \left(\frac{1}{2}, 0 \right)$$

$$w_3(x,y) = \left(\frac{1}{2}x, \frac{1}{2}y \right) + \left(\frac{1}{4}, \frac{\sqrt{3}}{4} \right)$$

We have seen that a system of iterated functions always converges to the same point of $\mathcal{H}(X)$ regardless of the point with which the iteration begins. So, to obtain the Sierpinski triangle, we do not necessarily have to start with a triangle, we can start iterating with any compact non-empty set of IR^2, like for instance a "full" square.

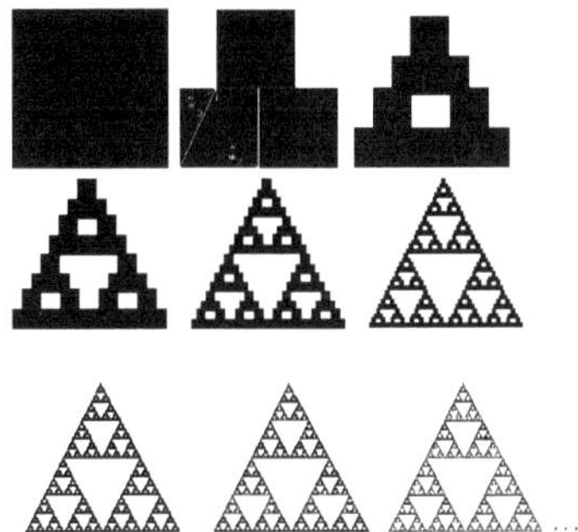

The fractal does not depend on the element of $\mathcal{H}(X)$ initially chosen

Some comments about the Sierpinski Triangle fractal:

The nth iteration is composed of 3^n triangles (if we start with a triangle; if we start with another figure, from N large enough, the difference between each of these figures composing the total figure and each of the triangles that would compose it if starting with a triangle, is irrelevant, given its tiny size) of side $(1/2)^n$ and its total area is $A_0 \times (3/4)^n$ where A_0 is the area of the initial triangle ($A_0 = \dfrac{\sqrt{3}}{4}$ if $l_0 = 1$), so the fractal has zero area:

$$\lim_{n \to \infty} A_0 \cdot \left(\frac{3}{4}\right)^n = A_0 \cdot \lim_{n \to \infty}\left(\frac{3}{4}\right)^n = A_0 \cdot 0 = 0$$

However, its perimeter is infinite, because

$$P_n = 3^n \times 3 \times \left(\frac{1}{2}\right)^n = 3 \times \left(\frac{3}{2}\right)^n \xrightarrow[n \to \infty]{} +\infty$$

6.4.2 - AN INTERVAL IN IR

IFS: $\{IR; w_1, w_2\}$ where:

$$w_1(x) = \frac{1}{3}x \qquad\qquad w_2(x) = \frac{2}{3}x + \frac{1}{3}$$

Shrinkage factor = max $\{1/3 \,; 2/3\} = 2/3$

Starting with $B_0 = [0, 1]$, let us determine B_1 :

$B_1 = W(B_0) = \tilde{w}_1 (B_0) \cup \tilde{w}_2 (B_0) = [0\,; 1/3] \cup [1/3\,; 1] = [0\,, 1] = B_0$

Then we will have $B_0 = B_1 = B_2 = \ldots = B_n = [0\,, 1] \xrightarrow[n \to \infty]{} [0\,, 1]$

Therefore, $[0\,, 1]$ is a fractal, although trivial, in IR.

6.4.3   - THE SINGER'S SET

IFS: $\{IR; w_1, w_2\}$ where:

$$w_1(x) = \frac{1}{3}x \qquad\qquad w_2(x) = \frac{1}{3}x + \frac{2}{3}$$

Shrinkage factor = 1/3

The only difference with the previous IFS is that the coefficients of the second contraction are swapped. It is enough to get a more interesting fractal.

Starting with $B_0 = [0, 1]$, let us determine B_1 :

$B_1 = W(B_0) = w_1(B_0) \cup w_2(B_0) = [0 ; 1/3] \cup [2/3 ; 1] = [0, 1] \setminus]1/3 ; 2/3[$

$B_2 = W(B_1) = \tilde{w}_1(B_1) \cup \tilde{w}_2(B_1) =$

$\quad = \tilde{w}_1([0 ; 1/3] \cup [2/3 ; 1]) \cup \tilde{w}_2([0 ; 1/3] \cup [2/3 ; 1]) =$

$\quad = \tilde{w}_1([0 ; 1/3]) \cup \tilde{w}_1([2/3 ; 1]) \cup \tilde{w}_2([0 ; 1/3]) \cup \tilde{w}_2([2/3 ; 1]) =$

$\quad = [0 ; 1/9] \cup [2/9 ; 1/3] \cup [2/3 ; 7/9] \cup [8/9 ; 1]$

And so on:

The term B_n of the succession consists of the disjoint meeting of 2^n closed intervals of amplitude $(1/3)^n$.

- The fractal "Cantor set" is defined by: $A = \bigcap_{n=0}^{\infty} B_n$.

- A is compact because it is defined to be the intersection of compact sets.

- A is totally disconnected (its only connected subsets are formed by a single point).

  A would not be totally disjoint if it contained at least one interval [a,b]. However, there is an order N from which a part of that interval is removed to form a term

B_{N+1} , so A does not contain intervals but only singular sets. It is also not empty - for example 1 and 0 belong to A.

- A is perfect (i.e. it has no isolated points - all its points, are points of accumulation).

Let x be$\in$ A and let$\varepsilon > 0$.

$$x \, A\in \Rightarrow x\in \bigcap_{n=0}^{\infty} B_n$$

$\forall$n, B_n consists of 2^n closed intervals In , In_{12} , ..., In $_2{}^n$

$\forall$n, k$\exists\in$ {1, 2, 3, ..., 2^n } : x$\in$ In $_k\subset$ B $._n$

Since the intervals In_k are closed, we can define two successions:

{m $\}_{nn\in IN}$ = {min {x$\in$ In }$\}_{kn\in IN}$ e

{M $\}_{nn\in IN}$ = {max {x$\in$ In }} $._{kn\in IN}$

So m $_n\leq$ x$\leq$ M , n N.$_n\forall\in$

Moreover, we know that each interval In_k has amplitude $(1/3)$ $^n \xrightarrow[n\to\infty]{}$ 0, so (M_n - m)$_n \xrightarrow[n\to\infty]{}$ 0, that is, both successions converge to x, at least one of them is not constant after a certain order.

We have, therefore, at least one succession of points of A convergent for x, not constant after a certain order, so that$\forall\varepsilon > 0$, V_ε (x) A$\cap\neq\varnothing$, that is, x is not an isolated point of A.

- Note the property of self-similarity (at any scale, i.e. at any of the iterations, the enlargement of a part is similar to one of the previous iterations).

6.4.4   - NON - LINEAR CASE

IFS:$\{IR; w_1, w_2\}$ where:

$$w_1(x) = \frac{1}{9}x^2 \qquad\qquad w_2(x) = \frac{3}{4}x + \frac{1}{2}$$

Shrinkage factor = max {4/9 ; 3/4} = ¾

In this case, one of the equations is of the second degree, which means that the open intervals eliminated in each iteration do not all have the same amplitude, but the fractal obtained has the same properties as the previous one.

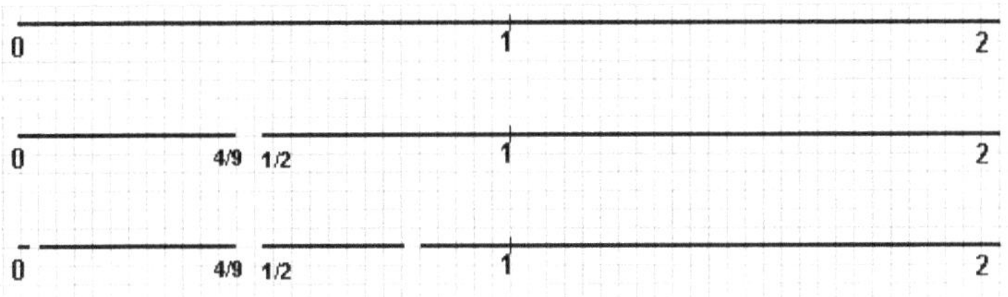

6.4.5 - IFS WITH CONDENSING SET IN THE COMPLEX PLANE

$$\text{IFS:}\left\{C;\ \tilde{w}_0,\ w_1,\ w_2\right\}\ \text{where:}$$

$$\forall B \in\ \mathcal{H}(X),\ \tilde{w}_0(B) = T \quad com \quad T = \left\{z \in C : -\frac{1}{2} \le \operatorname{Re}(z) \le \frac{1}{2}\ \wedge\ 0 \le \operatorname{Im}(z) \le 10\right\}$$

We say that w_0 is a condensation transformation and that T is the associated condensation set.

$$w_1(x) = 0{,}75e^{\frac{\pi}{4}i}z + 10i$$

$$w_2(x) = 0{,}75e^{-\frac{\pi}{4}i}z + 10i$$

Shrinkage factor = 0.75.

We could now modify some parameters in the functions and check what changes this causes in the image obtained.

Image obtained after five iterations.

This is one of the examples where we can observe the close relationship between Fractal Geometry and Nature. In fact, and contrary to Euclidean Geometry, which will hardly have in Nature a faithful representation of any of its objects, Fractal Geometry is perhaps the one that better adapts, in a more or less rough way (since, I say again, fractals in their pure state only exist in our minds) to the study of Nature's forms: trees and plants, snowflakes, blood irrigation systems and oxygen distribution, galaxies, cuttings of the coasts of continents, clouds, lightning, ... and also in a multitude of physical processes such as crystallisation, particle movements in a fluid, electrolysis, ...

6.4.6 - THE FRACTAL START

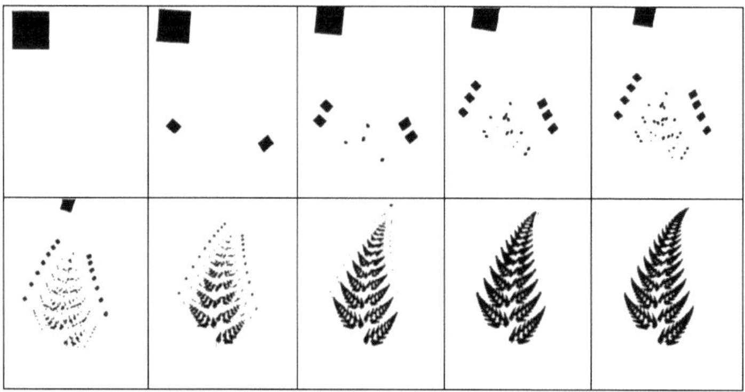

In an enlarged copy of the previous figures it is possible to determine the approximate value of the parameters that define the three applications that constitute the IFS that originates this fractal, very similar to the leaf of a fern. In this plant it is very easy to observe and understand the property of the self-similarity of a fractal since it grows repeating the same shape in smaller and smaller scales (a part of the leaf looks very much like the whole leaf).

The IFS on IR^2 is formed by three contractions, as from the first image to the second we observe that the square is transformed into three smaller quadrilaterals.

Each of the three contractions consists of a rescaling, with or without "distortion" followed by a translation. Thus, each of them can be written analytically in the form $w(x) = w\begin{pmatrix} x_1 \\ x_2 \end{pmatrix} = \begin{pmatrix} a & b \\ c & d \end{pmatrix}\begin{pmatrix} x_1 \\ x_2 \end{pmatrix} + \begin{pmatrix} e \\ f \end{pmatrix} = Ax + t$ and the matrix A can be written in the form $A = \begin{pmatrix} r_1\cos\theta_1 & -r_2\sin\theta_2 \\ r_1\sin\theta_1 & r_2\cos\theta_2 \end{pmatrix}$ where r_1 and r_2 correspond to the rescaling factors of each of the sides of the quadrilateral and θ_1 and θ_2 correspond to the distortion angles of each of these sides.

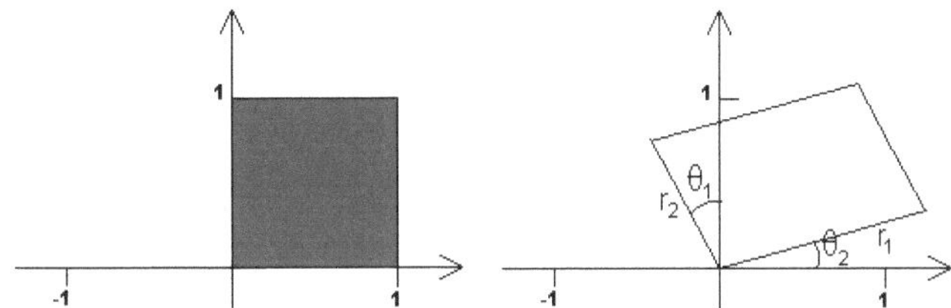

All that is needed now is to take some measurements with ruler and protractor, considering that the three quadrilaterals in the second figure are parallelograms, and then determine the parameters *a, b, c, d, e* and *f of* each of the three contractions.

Thus:

$$w_1\begin{pmatrix} x_1 \\ x_2 \end{pmatrix} = \begin{pmatrix} \dfrac{3}{3,25}\cos(-2) & -\dfrac{2,8}{3,25}\sin(-3) \\[2ex] \dfrac{3}{3,25}\sin(-2) & \dfrac{2,8}{3,25}\cos(-3) \end{pmatrix}\begin{pmatrix} x_1 \\ x_2 \end{pmatrix} + \begin{pmatrix} -0,5 \\ 0,75 \end{pmatrix}$$

$$w_1\begin{pmatrix} x_1 \\ x_2 \end{pmatrix} = \begin{pmatrix} \dfrac{0,95}{3,25}\cos(49) & -\dfrac{1,1}{3,25}\sin(49) \\[2ex] \dfrac{0,95}{3,25}\sin(49) & \dfrac{1,1}{3,25}\cos(49) \end{pmatrix}\begin{pmatrix} x_1 \\ x_2 \end{pmatrix} + \begin{pmatrix} 1,3 \\ -7 \end{pmatrix}$$

$$w_1\begin{pmatrix} x_1 \\ x_2 \end{pmatrix} = \begin{pmatrix} \dfrac{1}{3,25}\cos(-49) & -\dfrac{1,25}{3,25}\sin(-54) \\[2ex] \dfrac{1}{3,25}\sin(-49) & \dfrac{1,25}{3,25}\cos(-54) \end{pmatrix}\begin{pmatrix} x_1 \\ x_2 \end{pmatrix} + \begin{pmatrix} 8,4 \\ -7,95 \end{pmatrix}$$

6.4.7 - THE PEANO CURVE

This curve invented in 1890 by Giuseppe Peano is a fractal whose IFS consists of nine contractions of contraction factor 1/3 and whose parameters
are indicated alongside.

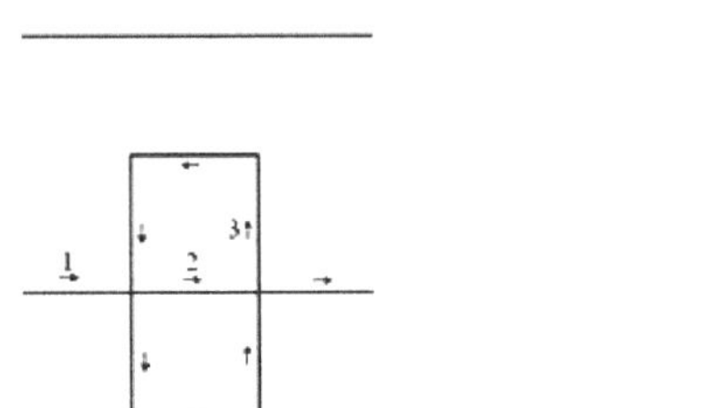

$$w\begin{pmatrix} x \\ y \end{pmatrix} = \begin{pmatrix} a & b \\ c & d \end{pmatrix}\begin{pmatrix} x \\ y \end{pmatrix} + \begin{pmatrix} e \\ f \end{pmatrix}$$

a	b	c	d	e	f
1/3	0	0	1/3	0	0
1/3	0	0	-1/3	0	-1/3
1/3	0	0	1/3	-1/3	-1/3
-1/3	0	0	1/3	1/3	-2/3
-1/3	0	0	-1/3	-2/3	0
-1/3	0	0	1/3	0	-1/3
1/3	0	0	1/3	-1/3	1/3
1/3	0	0	-1/3	1/3	-2/3
1/3	0	0	1/3	-2/3	0

To construct it we can start with a unit line segment, for example, and then place the nine segments of length 1/3 as shown in the figure on the left.

The result you get in the first few iterations is as follows:

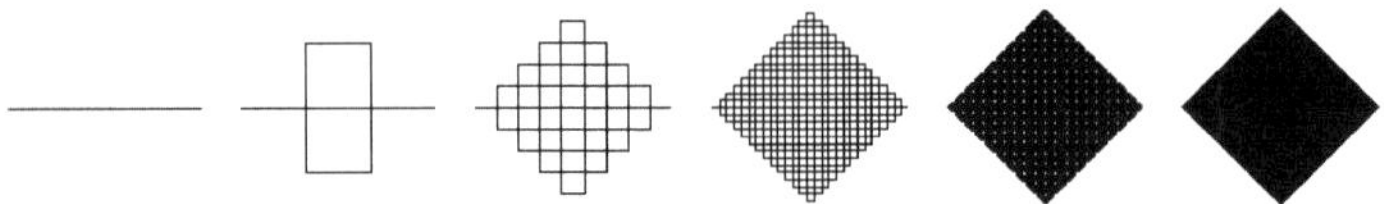

It proves that this fractal is a curve that completely covers a square.

So how big is this curve?

The concept of fractal dimension was created. It measures the "amount of plane, or space" that a fractal occupies or we can also think of it as a quantifier of its "roughness". Thus, the dimension of a curve can be any value in the interval [1 , 2].

7 - THE DIMENSION OF A FRACTAL

The dimension of a fractal is the value **D** such that $N = (1/r)^D$ where r is the contraction factor of the IFS that generates it and N is the total number of contractions that constitute the IFS.

So,

$$D = \frac{\log N}{\log\left(\frac{1}{r}\right)}.$$

For example, in the case of the Peano curve, we use nine segments (N=9) which are the reduction by a factor of 1/3 (r=1/3) of the previous segment. So D = log9/log3 = 2

We can only apply this construction of the notion of fractal dimension to fractals consisting of contractions of equal contraction factor.

Another process for determining the size of a fractal is "Box Counting" and consists of the following: first we start with an image which we suspect has fractal dimension (for example a photograph of a galaxy). Then we produce a binary version of that image: all pixels with a brightness above a certain chosen value are assigned the value one (we paint them white, for example) and the rest zero (we paint them black). Next we produce the image of the contours around the bright areas selected in the previous step.

Now we break the image into boxes of a certain side length and count how many of these boxes contain the outline. And then we repeat this process with different side lengths for the boxes.

As the number of squares covering the total area of the figure is proportional to the square of the inverse of the measure of the side of each square into which the figure was divided, it is now sufficient to use the results obtained in the previous measurements, and graph the logarithm of the number of squares containing the contour as a function of the logarithm of the inverse of the side of each square. The slope of the linear regression line that best fits the set of points obtained is the fractal dimension of the image. With either process, it is easy to verify that images with Euclidean dimension have equal value of fractal dimension.

7.1 - FRACTALS OF DIFFERENT DIMENSIONS

7.1.1 - The Koch Curve

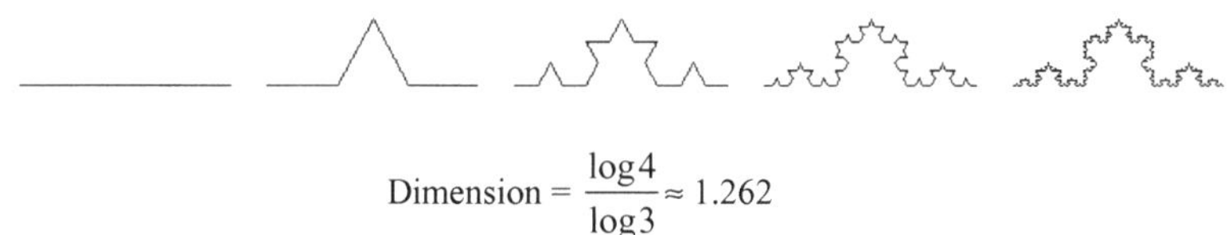

$$\text{Dimension} = \frac{\log 4}{\log 3} \approx 1.262$$

This curve has no derivative at any of its points.

7.1.2 - Peano Curves

The following curves are constructed in an identical manner to the Koch curve, by varying the reduction factor of each of the segments in relation to the initial segment and, automatically, the angle between these segments.

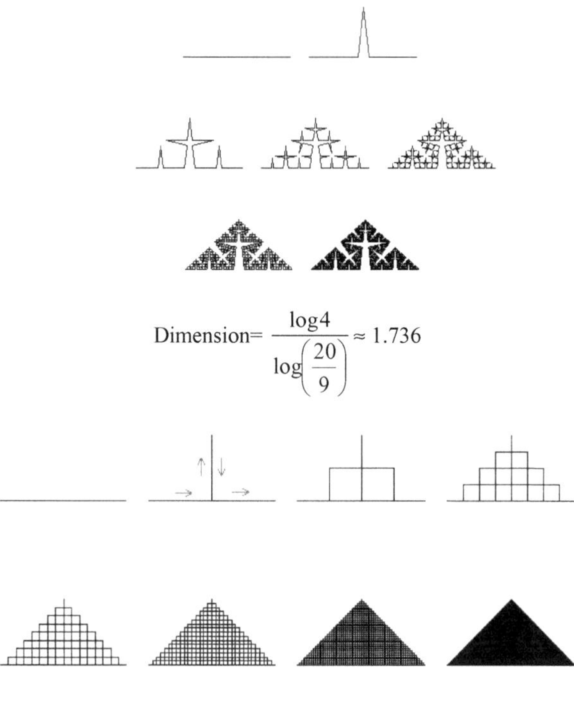

$$\text{Dimension} = \frac{\log 4}{\log\left(\frac{20}{9}\right)} \approx 1.736$$

$$\text{Dimension} = \frac{\log 4}{\log 2} = 2$$

7.1.3 - Sierpinski's Triangle

$$\text{Dimension} = \frac{\log 3}{\log 2} \approx 1.585$$

7.1.4 - The Sierpinski tetrahedron

A generalisation in space of the previous fractal

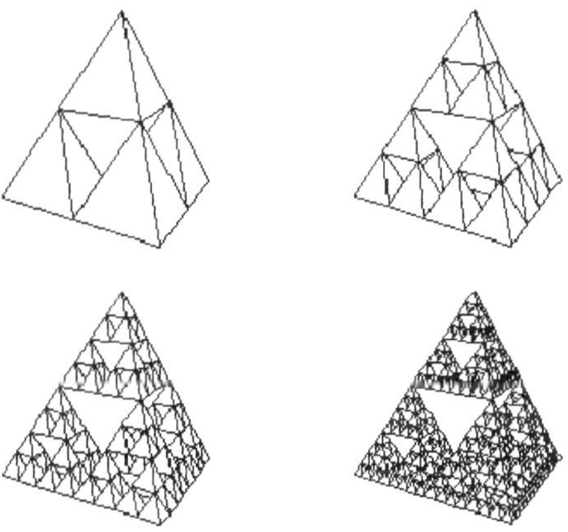

$$\text{Dimension} = \frac{\log 4}{\log 2} = 2$$

In this case we have an object in three-space, dimension 2. In each iteration, the total area remains constant as the volume tends towards zero.

7.1.5 - EXAMPLE OF THE APPLICABILITY OF THE CONCEPTS OF FRACTAL STRUCTURE AND DIMENSION

This concept of fractal dimension has had applicability in the most diverse fields, from mineralogy (to measure mineral density, terrain evolution, discontinuity in rocks), in geography (measuring the length of the coastline of continents), in biology (analysing the roughness of fungi and corals), in industry (for example, in the automatic detection of textile faults), etc, etc...

Next I present in a little more detail an example where the perception of the concepts of fractal structure and dimension were important for the development of science, namely biology and ecology. This example is taken from an article by Nigel Williams in the April 97 issue of *Science* magazine, where he relates the work of two ecologists - James Brown and Brian Enquist - and a physicist - Geoffrey West - who formed a team and combined their knowledge of the dynamics of energy transport in fluids (hydrodynamics) and mathematics in fractal geometry to try to solve a problem that intrigued biologists and ecologists.

Living organisms vary greatly in size, from microscopic creatures to whales, and biologists have long tried to understand the relationship between size and physiology and each living thing.

It was known from results obtained in various measurements that the metabolic rate is proportional to a power of exponent ¾ of the mass of the organism - the larger the creature, the slower its metabolism. And identical relations have been found for the increase in the population of a species, age at first reproduction, duration of embryo development, related to powers of the mass of exponents ¼, ¾ and -¼ respectively. The common thing in all these relations and which seems to be valid for all living organisms of the most diverse sizes, whether of the animal or vegetable kingdom, is the power of exponent ¼.

$$Y = \alpha M^{\frac{p}{4}}$$

For a long time, this factor astonished researchers, since, in the case of three-dimensional bodies, it would be much more logical for the exponent 1/3 to appear in the power.

This team analysed in geometrical and physical terms the linear systems of tubes that distribute resources (oxygen, food) and waste throughout the organism and considered that such systems would have to have three characteristics:

- The distribution network has to reach all points of the three-dimensional body,

- It should require minimal energy to transport these elements in a fluid medium,

- The last "little tubes" in the network (for example, the capillaries in a circulatory system) must all be the same size because cells in all living things are roughly the same size.

The answer came when the team realised that such a distribution network was best characterised by a fractal branching system for space filling.

8 - JULIA AND MANDELBROT'S SETS

8.1 - THE JULIA SET

Pierre Fatou (1878-1929) and Gaston Julia (1893-1978) used iterative methods it is possible to study fractals. Given the functions $x \rightarrow f(x) = z^2 - \mu \mid \mu = \lambda,$ $4 - \lambda, w$ e $z \rightarrow f(x) = z^2 - \mu \mid z = x + iy \wedge z, i \in I$ the technique consists in generating the graph for a given $\lambda \in IR$ and then this step is written in a tree and the process is repeated indefinitely.

The iterative method consists of the following steps, given a µ each iteration keeps µ fixed and modifies the value of Z as explained in [MATH3] and [MANDELBROT] page 183.

Thus a Julia set is a set of points, generated by each new iteration, that infinitely approach each other but never touch.

8.1.1 - Mandelbrot sets

Bertold B. Mandelbrot decided to redo all Julia's studies because he did not like the studies that Julia had done. When Mandelbrot finished his research he reached the same conclusion that Gaston Julia had reached. As it was a simpler model than Julia's and because he had computing resources, Mandelbrot managed to make his work the cradle of fractal theory and it was also possible to find a relationship between the work of other researchers (Koch, Julia, Cantor, among others), and his own.

Mandelbrot sets is the domain of convergence of the series of complex numbers obtained from the equation $Z_{n+1} = z_n^2 + C$. Where the variable C remains constant, however Z varies during the process. As we interact with Z, two things can happen: Either $Z \leq 2$ for any given value or $Z > 2$. If Z exceeds the value 2, the number C is not part of the Mandelbrot set. In other words, if this point remains near the origin ($Z < 2$) it belongs to the Mandelbrot set, if it leaves the vicinity of the origin and tends to infinity it does not belong to the set. It is possible to know if a given number C belongs or not to this set by checking its magnitude, its distance to the origin.

To obtain the magnitude of a given point by applying the Pythagorean theorem:

$$D = \sqrt{A^2 + B^2}$$ where D is the magnitude.

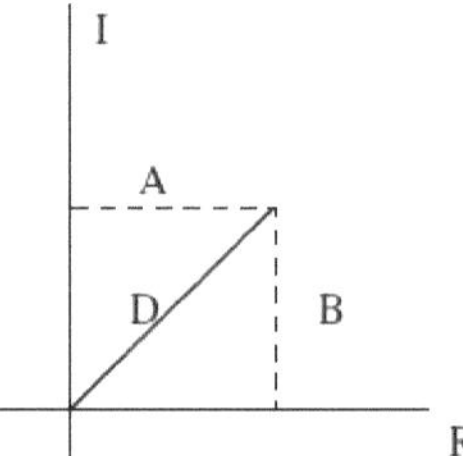

Figure 3.9: Line of imaginaries at y, line of reals at x, and D is the magnitude

In [MANDELBROT] Mandelbrot defines his set as follows: "Mandelbrot sets mark the points in the complex plane for which Julia's set is connected (can be separated into two subsets having at least one element), and non-computable".

By plotting a sample of millions of points of a number from Mandelbrot's set we obtain:

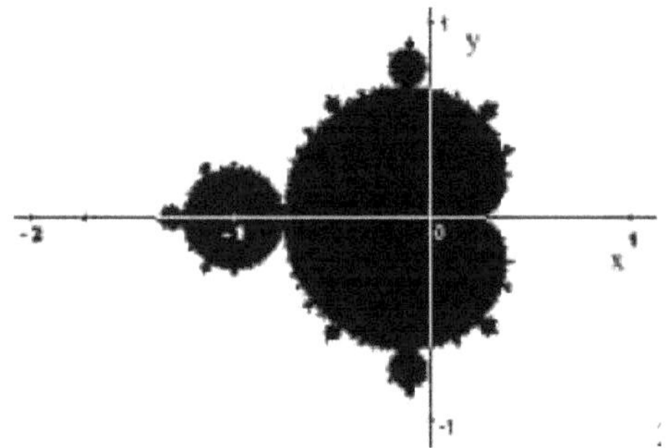

Figure 4: A sample when millions of points are plotted.

The complete structure of this set can be appreciated by expanding the different regions consecutively as shown:

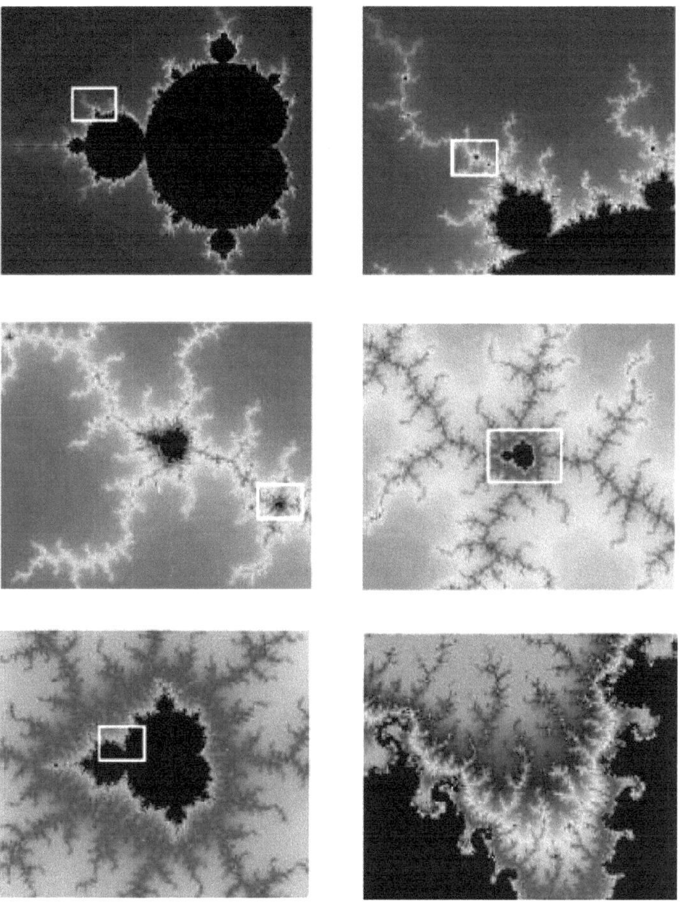

Figure 4.1: Fractal generated by *Ultra Fractal* software, where certain regions are

In the sequence of enlargements we can observe the self-similarity structure of this set (it contains copies of itself at smaller scales).

The following shows in the possible ways of programming a command to represent the Mandelbrot set: using the C language and using Mathematica software (see Wolfram, 1996) to simulate animations.

8.1.2 - Implementation of the Mandelbrot set in C language

To implement an algorithm to represent the Mandelbrot set we just need to answer the following question: How do we know if the orbit from 0 for a given value of c really tends to infinity? The trend criterion tells us that an orbit diverges if it ever leaves the circle of radius 2 centred at the origin. Writing in C Language, we can generate the following algorithm:

```c
1. #include <graphics.h>
2. #include <conio.h>
3. void main()
4. {
5. int sx=320; //Screenwidth
6. int sy=240; //Screenheight
7. double xmin=-2; //smallest real value (x-axis)
8. double xmax=1.25; //largest real value (x-axis)
9. double ymin=-1.25; //smallest imaginary value (y-axis)
10.        double ymax=1.25; //largest imaginary value (y-axis)
11.        int maxiter=96; //Max number of iterations
12.        double old_x; //temporary variable to store x-value
13.        double fx,fy;
14.        int m; //variable to store number of iterations
15.        int gd=VGA,gm=2; //MS-DOS spesific graphics-init.
16.        double dx=(xmax-xmin)/sx; //how much to add for each x-pixel?
17.        double dy=(ymax-ymin)/sy; //how much to add for each y-pixel?
18.        int px; //Variable storing current x-pixel
19.        int py=0; //Variable storing current y-pixel
20.        double x; //Variable storing current x-value
21.        double y=ymin; //Variable storing current y-value
22.        initgraph(&gd,&gm, "c:³³"); //code. Update the path!!!
23.        while (py<sy) {
24.        px=0;
25.        x=xmin;
26.        py++;
27.        while (px<sx) {
28.        px++;
29.        fx=0;
30.        fy=0;
31.        m=0;
32.        do {
33.        old_x=fx;
34.        fx=fx*fx-fy*fy+x;
35.        fy=2*old_x*fy+y;
36.        m++;
37.        } while ((((fx*fx+fy*fy)<4) && (m<maxiter));
38.        putpixel(px,py,m); //MS-DOS spesific code to give pixel(x,y)
39.        //the colour (m)
40.        x+=dx;
41.        }
42.        y+=dy;
43.        }
44.        getch();
45.        closegraph(); //MS-DOS spesific code to close
46.        graphicscreen
47.        }
```

Where this is the program that has been used to obtain the previous graphics, (variables sx and sy), the maximum and minimum values of the real parts (xmin and xmax) and the imaginary parts (ymin and ymax), as well as the maximum number of iterations performed to check whether the orbit converges or diverges (*maxiter*). Note that the greater this last value, the better the accuracy of the graph.

8.1.3 - Implementation of the Mandelbrot set with "Mathematica" software

The implementation of the previous algorithm with *Mathematica* software is much more sensitive and intuitive. See the code that generates the Mandelbrot fractal in 3D:

```
mandel = Compile[{{n,_Integer},{c0,_Complex},{r,_Real},
{lim,_Integer},{i,_Integer},{j,_Integer}},
Module[{dx=r/(n-1),dy=r/(n-1),z=I,c=I,k=1},
z=0; c=c0+(2*i-1-n)*dx+(2*j-1-n)*I*dy; k=0;
While[Abs[z]<2 && k<lim, z=z^2+c; k++]; k] ];
n=50; ans=Table[mandel[n, -0.5+.0*I, 1.5, 60, i, j],{i,1,n}, {j,1,n}];
ListPlot3D[ans, ColorFunction->Hue, ViewPoint->{1.317, -1.859, 2.502}]
```

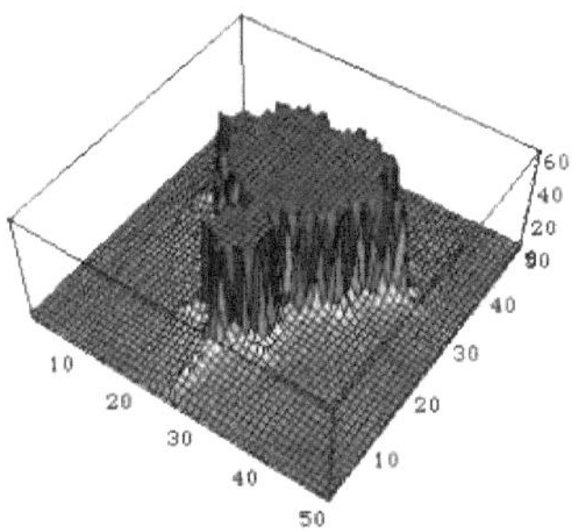

Figure 5: Mandelbrot fractal in 3D

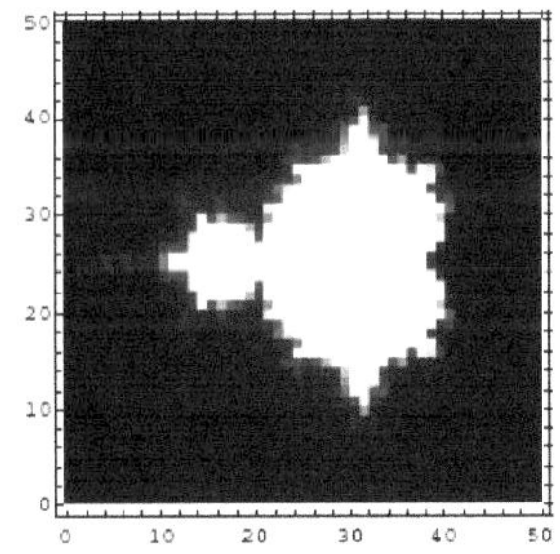

Figure 6: Mandelbrot fractal in 2D

9 - JULIA'S SET

To calculate the Mandelbrot set we were only interested in the orbit of 0 for each value of c. For the Julia set, we set a value of c and then consider the fate of all possible similarities for this fixed value of c. Those similar curves in orbits do not escape, form the Julia set. Therefore, we will obtain a different Julia set for each different c. The following figure shows the different Julia sets for different values of c:

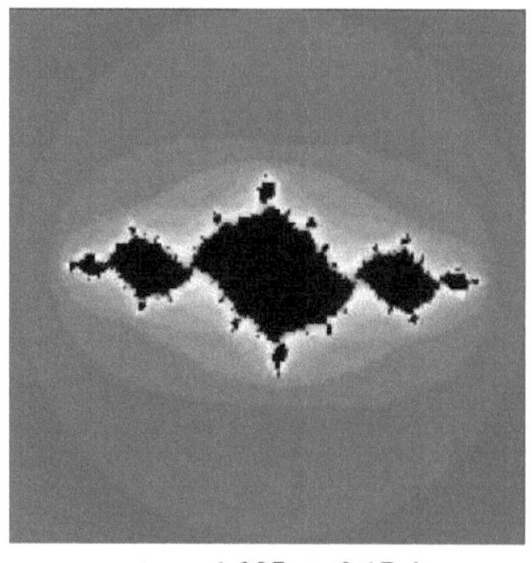

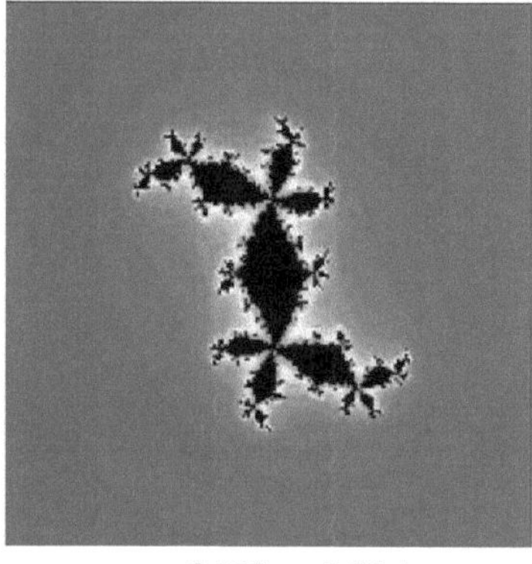

c = -1.037 + 0.17 i   c=0.295 + 0.55 i

One of the most surprising theorems of fractal geometry dates from 1919 and was proved by Gastón Julia and Pierre Fatou. The fundamental dichotomy for the equation $z^2 + c$ is: for each value of c, the Julia set is a connected set (one piece) a completely disconnected set (infinitely many pieces). Surprisingly, it is the orbit of 0 that determines whether a Julia set is connected or not: if the orbit of 0 tends to infinity then the Julia set is a nonconnected set, and if the orbit of 0 does not tend to infinity, then it is a connected set. Therefore, if we take values of c inside the Mandelbrot set, we tend to the set of Julia connected, and otherwise, the set will be totally disconnected.

9.1 - SOME ALGORITHMS THAT GENERATE JULIA FRACTALS AND THEIR RESPECTIVE IMAGES AS A RESULT OF PASCAL LANGUAGE

```
1.   Program Fractal_1;
2.   Function Pow (Num, Exponent: Integer): Integer;
3.   Begin
4.   If (Exponent=0) Then
5.   Pow:=1
6.   Else
7.   Pow:=Num*Pow (Num, Exponent-1);
8.   End;

9.   Var
10.  { variables for the fractal }
```

11. max_int, max_color, nx, ny, screen_x_max, screen_y_max, colour: Integer;
12. dx, dy, xx, yy, r, p, q, xmin, xmax, ymin, ymax, x0, y0, x, y: Real;
13. { variables for graphical mode }
14. max_num_color, gd, gm: Integer;
15. m: char;
16. output:Text;
17. { variables for compression }
18. Color_Ant, Cont, Posic, Driver, Mode: Integer;
19. Bit, Test: Byte;
20. Colours, Repeats: Array [1..8] of Byte;
21. Save: Boolean;

22. Begin
23. max_num_color:=255;
24. screen_x_max:=640;
25. screen_y_max:=480;
26. assign (output, 'JULIA1.FRC');
27. rewrite (output);

28. Driver:=9;
29. Mode:=2;

30. max_int:=250;
31. max_color:=250;
32. xmin:=-1.0;
33. xmax:=0.5;
34. ymin:=-1.0;
35. ymax:=0.5;
36. dx:=(xmax-xmin)/(screen_x_max-1);
37. dy:=(ymax-ymin)/(screen_y_max-1);
38. p:=0.39054;
39. q:=-0.58679;
40. nx:=-1;
41. ny:=-1;

42. cor_ant:=-1;
43. bit:=0;
44. Posic:=0;
45. record:=False;

46. nx:=-1;
47. While (nx<=max_screen) do Begin
48. Inc (nx);
49. ny:=-1;
50. While (ny<=Y_Max_Screen) do Begin
51. Inc (ny);
52. x0:=xmin+nx*dx;
53. y0:=ymin+ny*dy;
54. colour:=0;
55. x:=x0;
56. y:=y0;
57. r:=0;
58. While (r < max_int) and (colour < max_color) do Begin
59. xx:=x*x-y*y+p;
60. yy:=2*x*y+q;
61. r:=xx*xx+yy*yy;

```
62.  inc (colour);
63.  x:=xx;
64.  y:=yy;
65.  End;
66.  colour:=colour mod max_num_colour;
67.  m:=chr (colour);
68.  if (Ant_Colour=Colour) Then
69.  if (Repeats [Position]=254) Then Begin
70.  Inc (Repeats [Posic]);
71.  Ant_colour:=-1;
72.  End
73.  Else Begin
74.  Inc (Repeats [Posic]);
75.  Bit:=Bit or pow (2, Posic-1);
76.  End
77.  Else Begin
78.  Inc (Posic);
79.  If (Posic=9) Then
80.  Save:=True
81.  Else Begin
82.  Repeats [Position]:=0;
83.  Colours [Posic]:=Colour;
84.  Ant_Colour:=Colour;
85.  End
86.  End;
87.  If (Save) Then Begin
88.  Write (Output, Chr (Bit));
89.  For Cont:=1 to 8 do Begin
90.  Test:=Pow (2, Cont-1);
91.  If ((Bit and Test)=Test) Then
92.  Write (Output, Chr (Repeats [Cont]));
93.  Write (Output, Chr (Colors [Cont]));
94.  End;
95.  Bit:=0;
96.  Posic:=1;
97.  Repeats [Position]:=0;
98.  Colours [Posic]:=Colour;
99.  Save:=False;
100.Ant_Colour:=Colour;
101.End
102.End
103.End;
104.If (Position>0) Then Begin
105.Write (Output, Chr (Bit));
106.For Cont:=1 to Posic do Begin
107.Test:=Pow (2, Cont-1);
108.If ((Bit and Test)=Test) Then
109.Write (Output, Chr (Repeats [Cont]));
110.Write (Output, Chr (Colors [Cont]));
111.End;
112.End;
113.Close (Exit);
114.End.
```

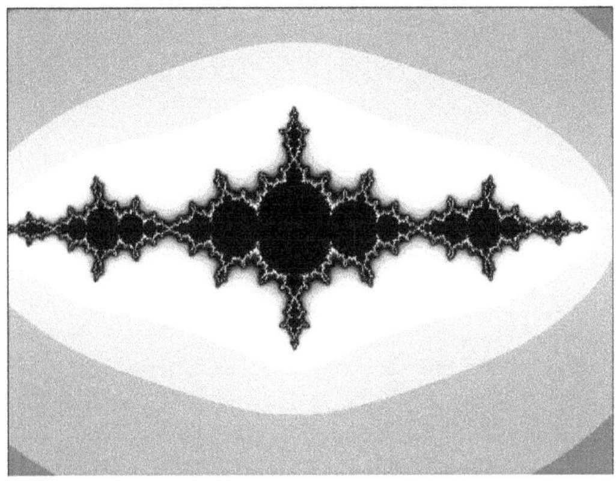

Figure 7 - Julia's Fractal

1. Program Fractal_2;
2. Function Pow (Num, Exponent: Integer): Integer;
3. Begin
4. If (Exponent=0) Then
5. Pow:=1
6. Else
7. Pow:=Num*Pow (Num, Exponent-1);
8. End;

9. Var
10. { variables for the fractal }
11. max_int, max_color, nx, ny, screen_x_max, screen_y_max, colour: Integer;
12. dx, dy, xx, yy, r, p, q, xmin, xmax, ymin, ymax, x0, y0, x, y: Real;
13. { variables for graphical mode }
14. max_num_color, gd, gm: Integer;
15. m: char;
16. output:Text;
17. { variables for compression }
18. Color_Ant, Cont, Posic, Driver, Mode: Integer;
19. Bit, Test: Byte;
20. Colours, Repeats: Array [1..8] of Byte;
21. Save: Boolean;

22. Begin
23. Driver:=9;
24. Mode:=2;
25. max_num_color:=256;
26. screen_x_max:=640;
27. screen_y_max:=480;
28. assign (output, 'JULIA2.FRC');
29. rewrite (output);

30. max_int:=256;
31. max_color:=256;

97

```
32. xmin:=-1.0;
33. xmax:=0.5;
34. ymin:=-1.0;
35. ymax:=0.5;
36. dx:=(xmax-xmin)/(screen_x_max-1);
37. dy:=(ymax-ymin)/(screen_y_max-1);
38. p:=0.39054;
39. q:=-0.58679;
40. nx:=-1;
41. ny:=-1;

42. cor_ant:=-1;
43. bit:=0;
44. Posic:=0;
45. record:=False;

46. nx:=-1;
47. While (nx<=max_screen) do Begin
48. Inc (nx);
49. ny:=-1;
50. While (ny<=Y_Max_Screen) do Begin
51. Inc (ny);
52. x0:=xmin+nx*dx;
53. y0:=ymin+ny*dy;
54. colour:=0;
55. x:=x0;
56. y:=y0;
57. r:=0;
58. While (r < max_int) and (colour < max_color) do Begin
59. xx:=x*x+y+p;
60. yy:=2*x*y+q;
61. r:=xx*xx+yy*yy;
62. inc (colour);
63. x:=xx;
64. y:=yy;
65. End;
66. colour:=colour mod max_num_colour;
67. m:=chr (colour);
68. if (Color_Ant=Color) Then
69. if (Repeats [Position]=254) Then Begin
70. Inc (Repeats [Posic]);
71. Ant_colour:=-1;
72. End
73. Else Begin
74. Inc (Repeats [Posic]);
75. Bit:=Bit or pow (2, Posic-1);
76. End
77. Else Begin
78. Inc (Posic);
79. If (Posic=9) Then
80. Save:=True
81. Else Begin
82. Repeats [Position]:=0;
83. Colours [Posic]:=Colour;
84. Ant_Colour:=Colour;
```

85. End
86. End;
87. If (Save) Then Begin
88. Write (Output, Chr (Bit));
89. For Cont:=1 to 8 do Begin
90. Test:=Pow (2, Cont-1);
91. If ((Bit and Test)=Test) Then
92. Write (Output, Chr (Repeats [Cont]));
93. Write (Output, Chr (Colors [Cont]));
94. End;
95. Bit:=0;
96. Posic:=1;
97. Repeats [Position]:=0;
98. Colours [Posic]:=Colour;
99. Save:=False;
100.Ant_Colour:=Colour;
101.End
102.End
103.End;
104.If (Position>0) Then Begin
105.Write (Output, Chr (Bit));
106.For Cont:=1 to Posic do Begin
107.Test:=Pow (2, Cont-1);
108.If ((Bit and Test)=Test) Then
109.Write (Output, Chr (Repeats [Cont]));
110.Write (Output, Chr (Colors [Cont]));
111.End;
112.End;
113.Close (Exit);
114.End.

Figure 7.1 - Julia Fractal

1.   Program Fractal_3;

2.   Function Pow (Num, Exponent: Integer): Integer;
3.   Begin
4.   If (Exponent=0) Then
5.   Pow:=1
6.   Else
7.   Pow:=Num*Pow (Num, Exponent-1);
8.   End;

9.   Var
10. { variables for the fractal }
11. max_int, max_color, nx, ny, screen_x_max, screen_y_max, colour: Integer;
12. dx, dy, xx, yy, r, p, q, xmin, xmax, ymin, ymax, x0, y0, x, y: Real;
13. { variables for graphical mode }
14. max_num_color, gd, gm: Integer;
15. m: char;
16. output:Text;
17. { variables for compression }
18. Color_Ant, Cont, Posic,Driver, Mode: Integer;
19. Bit, Test: Byte;
20. Colours, Repeats: Array [1..8] of Byte;
21. Save: Boolean;

22. Begin
23. max_num_color:=255;
24. screen_x_max:=640;
25. screen_y_max:=480;
26. assign (output, 'JULIA4.FRC');
27. rewrite (output);

28. Driver:=9;
29. Mode:=2;
30. max_int:=250;
31. max_color:=250;
32. xmin:=-1.6;
33. xmax:=1.6;
34. ymin:=-1.5;
35. ymax:=1.5;
36. dx:=(xmax-xmin)/(screen_x_max-1);
37. dy:=(ymax-ymin)/(screen_y_max-1);
38. p:=0.32;
39. q:=0.043;
40. nx:=-1;
41. ny:=-1;

42. cor_ant:=-1;
43. bit:=0;
44. Posic:=0;
45. record:=False;

46. nx:=-1;
47. While (nx<=max_screen) do Begin
48. Inc (nx);
49. ny:=-1;

50. While (ny<=Y_Max_Screen) do Begin
51. Inc (ny);
52. x0:=xmin+nx*dx;
53. y0:=ymin+ny*dy;
54. colour:=0;
55. x:=x0;
56. y:=y0;
57. r:=0;
58. While (r < max_int) and (colour < max_color) do Begin
59. xx:=x*x-y*y+p;
60. yy:=2*x*y+q;
61. r:=xx*xx+yy*yy;
62. inc (colour);
63. x:=xx;
64. y:=yy;
65. End;
66. colour:=colour mod max_num_colour;
67. m:=chr (colour);
68. if (Color_Ant=Color) Then
69. if (Repeats [Position]=254) Then Begin
70. Inc (Repeats [Posic]);
71. Ant_colour:=-1;
72. End
73. Else Begin
74. Inc (Repeats [Posic]);
75. Bit:=Bit or pow (2, Posic-1);
76. End
77. Else Begin
78. Inc (Posic);
79. If (Posic=9) Then
80. Save:=True
81. Else Begin
82. Repeats [Position]:=0;
83. Colours [Posic]:=Colour;
84. Ant_Colour:=Colour;
85. End
86. End;
87. If (Save) Then Begin
88. Write (Output, Chr (Bit)),
89. For Cont:=1 to 8 do Begin
90. Test:=Pow (2, Cont-1);
91. If ((Bit and Test)=Test) Then
92. Write (Output, Chr (Repeats [Cont]));
93. Write (Output, Chr (Colors [Cont]));
94. End;
95. Bit:=0;
96. Posic:=1;
97. Repeats [Position]:=0;
98. Colours [Posic]:=Colour;
99. Save:=False;
100.Ant_Colour:=Colour;
101.End
102.End
103.End;
104.If (Position>0) Then Begin

105. Write (Output, Chr (Bit));
106. For Cont:=1 to Posic do Begin
107. Test:=Pow (2, Cont-1);
108. If ((Bit and Test)=Test) Then
109. Write (Output, Chr (Repeats [Cont]));
110. Write (Output, Chr (Colors [Cont]));
111. End;
112. End;
113. Close (Exit);
114. End.

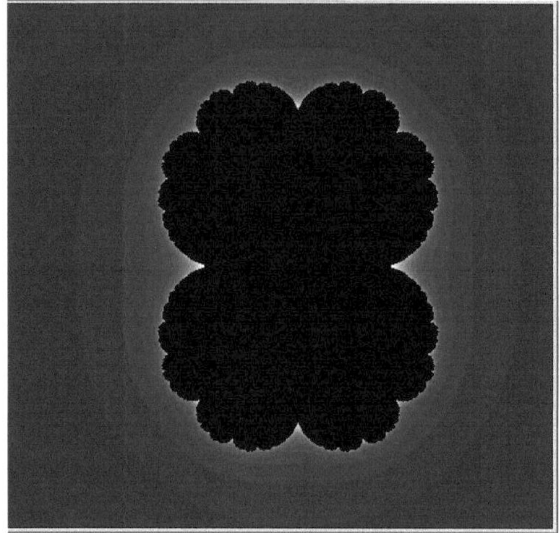

Figure 7.2 - Julia Fractal

9.2 - SOME ALGORITHMS THAT GENERATE MANDELBROT FRACTALS AND THEIR RESPECTIVE IMAGES DEVELOPED IN PASCAL LANGUAGE

1. Program Fractal_1;

2. Function Pow (Num, Exponent: Integer): Integer;
3. Begin
4. If (Exponent=0) Then
5. Pow:=1
6. Else
7. Pow:=Num*Pow (Num, Exponent-1);
8. End;

9. Var
10. { variables for the fractal }
11. max_int, max_color, nx, ny, screen_x_max, screen_y_max, colour: Integer;
12. dx, dy, xx, yy, r, p, q, xmin, xmax, ymin, ymax, x0, y0, x, y,

```
13. pmin, pmax, qmin, qmax, p0, q0: Real;
14. { variables for graphical mode }
15. max_num_color, gd, gm: Integer;
16. m: char;
17. output:Text;
18. { Variables for compression }
19. Col_Ant, Cont, Posic: Integer;
20. Bit, Test: Byte;
21. Colours, Repeats: Array [1..8] of Byte;
22. Save: Boolean;

23. Begin
24. screen_x_max:=640;
25. screen_y_max:=480;
26. assign (output, 'MANDEL1.FRC');
27. rewrite (output);

28. max_int:=100;
29. max_color:=256;
30. xmin:=-2.25;
31. xmax:=0.75;
32. ymin:=-1.5;
33. ymax:=1.5;
34. nx:=-1;
35. ny:=-1;

36. pmin := -2.25;
37. pmax := 0.75;
38. qmin := -1.5;
39. qmax := 1.5;

40. dx:=(pmax-pmin)/(screen_x_max-1);
41. dy:=(qmax-qmin)/(screen_y_max-1);

42. cor_ant:=-1;
43. bit:=0;
44. Posic:=0;
45. record:=False;

46. nx:=-1;
47. While (nx<=max_screen) do Begin
48. Inc (nx);
49. ny:=-1;
50. While (ny<=Y_Max_Screen) do Begin
51. Inc (ny);
52. p:=pmin+nx*dx;
53. q:=qmin+ny*dy;
54. colour:=0;
55. x:=0;
56. y:=0;
57. r:=0;
58. While (r <= max_int) and (colour < max_color) do Begin
59. xx:=x*x-y*y+p;
60. yy:=2*x*y+q;
61. r:=xx*xx+yy*yy;
```

62. inc (colour);
63. x:=xx;
64. y:=yy;
65. End;
66. colour:=colour mod max_colour;
67. m:=chr (colour);
68. if (Color_Ant=Color) Then
69. if (Repeats [Position]=254) Then Begin
70. Inc (Repeats [Posic]);
71. Ant_colour:=-1;
72. End
73. Else Begin
74. Inc (Repeats [Posic]);
75. Bit:=Bit or pow (2, Posic-1);
76. End
77. Else Begin
78. Inc (Posic);
79. If (Posic=9) Then
80. Save:=True
81. Else Begin
82. Repeats [Position]:=0;
83. Colours [Posic]:=Colour;
84. Ant_Colour:=Colour;
85. End
86. End;
87. If (Save) Then Begin
88. Write (Output, Chr (Bit));
89. For Cont:=1 to 8 do Begin
90. Test:=Pow (2, Cont-1);
91. If ((Bit and Test)=Test) Then
92. Write (Output, Chr (Repeats [Cont]));
93. Write (Output, Chr (Colors [Cont]));
94. End;
95. Bit:=0;
96. Posic:=1;
97. Repeats [Position]:=0;
98. Colours [Posic]:=Colour;
99. Save:=False;
100.Ant_Colour:=Colour;
101.End
102.End
103.End;
104.If (Position>0) Then Begin
105.Write (Output, Chr (Bit));
106.For Cont:=1 to Posic do Begin
107.Test:=Pow (2, Cont-1);
108.If ((Bit and Test)=Test) Then
109.Write (Output, Chr (Repeats [Cont]));
110.Write (Output, Chr (Colors [Cont]));
111.End;
112.End;
113.Close (Exit);
114.End.

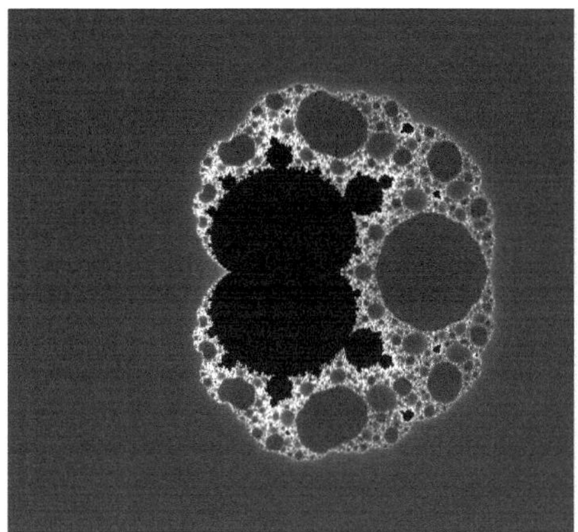

Figure 7.3 - Mandelbrot Fractal

1. Program Fractal_2;
2. Function Pow (Num, Exponent: Integer): Integer;
3. Begin
4. If (Exponent=0) Then
5. Pow:=1
6. Else
7. Pow:=Num*Pow (Num, Exponent-1);
8. End;
9. Var
10. { variables for the fractal }
11. max_int, max_color, nx, ny, screen_x_max, screen_y_max, colour: Integer;
12. dx, dy, xx, yy, r, p, q, xmin, xmax, ymin, ymax, x0, y0, x, y,
13. pmin, pmax, qmin, qmax, p0, q0: Real;
14. { variables for graphical mode }
15. max_num_color, gd, gm: Integer;
16. m: char;
17. output:Text;
18. { variables for compression }
19. Col_Ant, Cont, Posic: Integer;
20. Bit, Test: Byte;
21. Colours, Repeats: Array [1..8] of Byte;
22. Save: Boolean;
23. Begin
24. screen_x_max:=640;
25. screen_y_max:=480;
26. assign (output, 'MANDEL2.FRC');
27. rewrite (output);
28. max_int:=100;
29. max_color:=256;
30. xmin:=-2.25;
31. xmax:=0.75;
32. ymin:=-1.5;
33. ymax:=1.5;
34. pmin := -0.19920;
35. pmax := -0.12954;
36. qmin := 1.01480;
37. qmax := 1.06707;

105

```
38.  dx:=(pmax-pmin)/(screen_x_max-1);
39.  dy:=(qmax-qmin)/(screen_y_max-1);
40.  nx:=-1;
41.  ny:=-1;
42.  cor_ant:=-1;
43.  bit:=0;
44.  Posic:=0;
45.  record:=False;
46.  nx:=-1;
47.  While (nx<=max_screen) do Begin
48.  Inc (nx);
49.  ny:=-1;
50.  While (ny<=Y_Max_Screen) do Begin
51.  Inc (ny);
52.  p:=pmin+nx*dx;
53.  q:=qmin+ny*dy;
54.  colour:=0;
55.  x:=0;
56.  y:=0;
57.  r:=0;
58.  While (r <= max_int) and (colour < max_color) do Begin
59.  xx:=x*x-y*y+p;
60.  yy:=2*x*y+q;
61.  r:=xx*xx+yy*yy;
62.  inc (colour);
63.  x:=xx;
64.  y:=yy;
65.  End;
66.  colour:=colour mod max_colour;
67.  m:=chr (colour);
68.  if (Color_Ant=Color) Then
69.  if (Repeats [Position]=254) Then Begin
70.  Inc (Repeats [Posic]);
71.  Ant_colour:=-1;
72.  End
73.  Else Begin
74.  Inc (Repeats [Posic]);
75.  Bit:=Bit or pow (2, Posic-1);
76.  End
77.  Else Begin
78.  Inc (Posic);
79.  If (Posic=9) Then
80.  Save:=True
81.  Else Begin
82.  Repeats [Position]:=0;
83.  Colours [Posic]:=Colour;
84.  Ant_Colour:=Colour;
85.  End
86.  End;
87.  If (Save) Then Begin
88.  Write (Output, Chr (Bit));
89.  For Cont:=1 to 8 do Begin
90.  Test:=Pow (2, Cont-1);
91.  If ((Bit and Test)=Test) Then
92.  Write (Output, Chr (Repeats [Cont]));
93.  Write (Output, Chr (Colors [Cont]));
94.  End;
95.  Bit:=0;
96.  Posic:=1;
97.  Repeats [Position]:=0;
98.  Colours [Posic]:=Colour;
```

99. Save:=False;
100. Ant_Colour:=Colour;
101. End
102. End
103. End;
104. If (Position>0) Then Begin
105. Write (Output, Chr (Bit));
106. For Cont:=1 to Posic do Begin
107. Test:=Pow (2, Cont-1);
108. If ((Bit and Test)=Test) Then
109. Write (Output, Chr (Repeats [Cont]));
110. Write (Output, Chr (Colors [Cont]));
111. End;
112. End;
113. Close (Exit);
114. End.

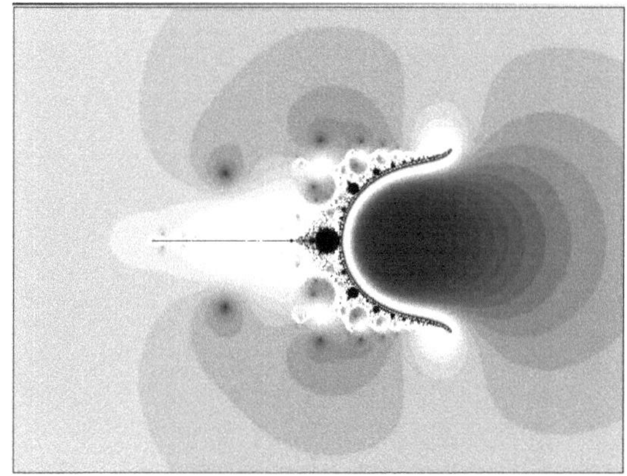

Figure 7.4 - Mandelbrot Fractal

1.   Program Fractal_3;
2.   Function Pow (Num, Exponent: Integer): Integer;
3.   Begin
4.   If (Exponent=0) Then
5.   Pow:=1
6.   Else
7.   Pow:=Num*Pow (Num, Exponent-1);
8.   End;
9.   Var
10. { variables for the fractal }
11. max_int, max_color, nx, ny, screen_x_max, screen_y_max, colour: Integer;
12. dx, dy, xx, yy, r, p, q, xmin, xmax, ymin, ymax, x0, y0, x, y,
13. pmin, pmax, qmin, qmax, p0, q0: Real;
14. { variables for graphical mode }
15. max_num_color, gd, gm: Integer;
16. m: char;
17. output:Text;
18. { variables for compression }
19. Col_Ant, Cont, Posic: Integer;
20. Bit, Test: Byte;

```pascal
21.  Colours, Repeats: Array [1..8] of Byte;
22.  Save: Boolean;
23.  Begin
24.  screen_x_max:=640;
25.  screen_y_max:=480;
26.  assign (output, 'MANDEL12.FRC');
27.  rewrite (output);
28.  max_int:=100;
29.  max_color:=256;
30.  xmin:=-2.25;
31.  xmax:=0.75;
32.  ymin:=-1.5;
33.  ymax:=1.5;
34.  pmin := -1.08621875;
35.  pmax := -1.0618125;
36.  qmin := 0.26598;
37.  qmax := 0.29076;
38.  dx:=(pmax-pmin)/(screen_x_max-1);
39.  dy:=(qmax-qmin)/(screen_y_max-1);
40.  nx:=-1;
41.  ny:=-1;

42.  cor_ant:=-1;
43.  bit:=0;
44.  Posic:=0;
45.  record:=False;

46.  nx:=-1;
47.  While (nx<=max_screen) do Begin
48.  Inc (nx);
49.  ny:=-1;
50.  While (ny<=Y_Max_Screen) do Begin
51.  Inc (ny);
52.  p:=pmin+nx*dx;
53.  q:=qmin+ny*dy;
54.  colour:=0;
55.  x:=0;
56.  y:=0;
57.  r:=0;
58.  While (r <= max_int) and (colour < max_color) do Begin
59.  xx:=x*x-y*y+p;
60.  yy:=2*x*y+q;
61.  r:=xx*xx+yy*yy;
62.  inc (colour);
63.  x:=xx;
64.  y:=yy;
65.  End;
66.  colour:=colour mod max_colour;
67.  m:=chr (colour);
68.  if (Ant_Colour=Colour) Then
69.  if (Repeats [Position]=254) Then Begin
70.  Inc (Repeats [Posic]);
71.  Ant_colour:=-1;
72.  End
73.  Else Begin
74.  Inc (Repeats [Posic]);
75.  Bit:=Bit or pow (2, Posic-1);
76.  End
77.  Else Begin
78.  Inc (Posic);
79.  If (Posic=9) Then
```

```
80. Save:=True
81. Else Begin
82. Repeats [Position]:=0;
83. Colours [Posic]:=Colour;
84. Ant_Colour:=Colour;
85. End
86. End;
87. If (Save) Then Begin
88. Write (Output, Chr (Bit));
89. For Cont:=1 to 8 do Begin
90. Test:=Pow (2, Cont-1);
91. If ((Bit and Test)=Test) Then
92. Write (Output, Chr (Repeats [Cont]));
93. Write (Output, Chr (Colors [Cont]));
94. End;
95. Bit:=0;
96. Posic:=1;
97. Repeats [Position]:=0;
98. Colours [Posic]:=Colour;
99. Save:=False;
100. Ant_Colour:=Colour;
101. End
102. End
103. End;
104. If (Posic>0) Then Begin
105. Write (Output, Chr (Bit));
106. For Cont:=1 to Posic do Begin
107. Test:=Pow (2, Cont-1);
108. If ((Bit and Test)=Test) Then
109. Write (Output, Chr (Repeats [Cont]));
110. Write (Output, Chr (Colors [Cont]));
111. End;
112. End;
113. Close (Exit);
114. End.
```

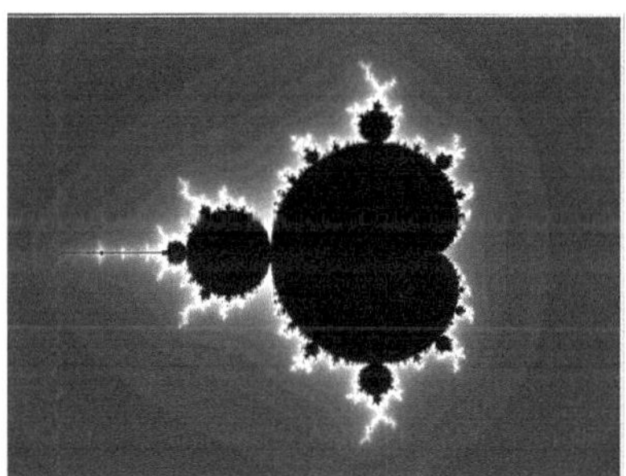

Figure 7.5 - Mandelbrot Fractal

10 - USING FRACTALS TO MODEL ELEMENTS OF NATURE

This topic discusses how to model some specific elements, applying the methods exposed in the previous sections and making use of some functions.

10.1 - LAND

Of the elements of nature to be addressed in this work, this is the simplest to model, since a terrain has a well-defined geometric surface. There are numerous ways to create terrain and relief models. In this work, it is appropriate to approach the procedural methods, or more specifically, using fractals and making use of the basic functions seen previously. It is worth pointing out that this method is not always the best solution, as will be seen later on. The geometric models that will be created here will be generated from a mesh of polygons. The spatial coordinates of each vertex will be the input parameters for the procedural functions and the result obtained will correspond to the height displacement value of the vertex in question.

The advantage of creating terrains procedurally consists mainly in the fact that one can have a relatively sophisticated parametric control with respect to the model. In addition, there are no problems concerning the resolution of the map that will be applied to the mesh. However, the main disadvantage of modelling a terrain in a procedural way is the fact that it is partly impossible to create a specific terrain (procedural methods create random patterns), and a digital image serving as a height map for the mesh is required for this purpose. A technique that tries to merge both techniques in order to acquire the advantages of each one is presented below.

The idea of modelling terrain procedurally was introduced by Mandelbrot [Mandelbrot 82], who observed a slight similarity of a curve generated by fBm* in the plane, with the outline of mountains. Performing an extension to three-dimensional space, he sought to create a surface similar to a mountain range. The result obtained resembled the model in Figure 1.0, which, as can be seen, still does not correspond exactly to the geometry of a mountain. This is because fractal functions that use iterations of some basic function usually result in statistically homogeneous and isotropic functions, when built in additive cascade. Thus, this type of function will not adequately represent a landform, because in nature land patterns are rarely uniformly spread out. A mountain range often rises from a plain, there are parts of mountains that are more eroded than others by the effect of erosion, there are valleys formed by rivers, etc.

Figure 8 - Terrain generated using the fBm function directly

* The function fBm *(Fractal Brownian Motion)* is a generalisation of the definition of fractals. First it is defined how the function for amplitude will be. In a different and more generic way, it is now sought to build a table with specific amplitude values for each iteration to be performed. For this table to be correct, these values must be decreasing and within the interval [0,1].

For this reason, one seeks to construct the fractal functions in multiplicative cascade, thus obtaining the multifractal functions. The most basic multifractal function consists only in changing the summation within the fractal iteration by a multiplication.

The effect of this new function is to create a heterogeneous spectrum. When applied to a mesh, it will produce regions that are flatter and others that are more rugged, resembling more closely the geographic features. Starting from this idea, several variations of the multifractal functions can be created, in order to simulate some phenomenon in the geological formation of the terrains, according to the impressionistic modelling method. Initially, it is observed that in some mountain formations, the plains, from which the mountains emerge, are regions more worn by erosion, and therefore more rounded, while the higher parts of the relief, have a greater noise, because they are geologically more recent. To create this kind of model, a coefficient can be added to the multifractal iteration, which will make the lower values (corresponding to the lower regions) more homogeneous, while for the higher values (corresponding to the higher regions) it is sought to make the noise more evident. One way of doing this consists in making a pre-evaluation of the point in question: if the point corresponds to a higher region, a coefficient close to one is used, which will maintain the original noise of the function. As the height values are lower, smaller and smaller coefficients are used, cancelling the original noise to some extent.

The pseudocode below shows how this can be done:

Height = function_Basic (P) // Pre-assessment of the height of point P on the terrain
For i = 1 to Octaves do
*increment = function_Basic(P) * Amplitude table[i] * Height;*
Height = Height + increment; // The increment coefficient tends not to change
a lot when the height is small.
*P = P * gap;*

Figure 8.1 is an example of a terrain created with this variation of the multifractal function.

Figure 8.1 - Terrain generated using multifractal function with noise damping coefficient inversely proportional to height.

A variation for this type of formation consists in not allowing the plateaus to have noise, regardless of the height at which they are located. To do this, the coefficient should be made to increase or decrease as a function of height variation, and no longer as a function of absolute height.

Figure 8.2 shows an implementation of this function.

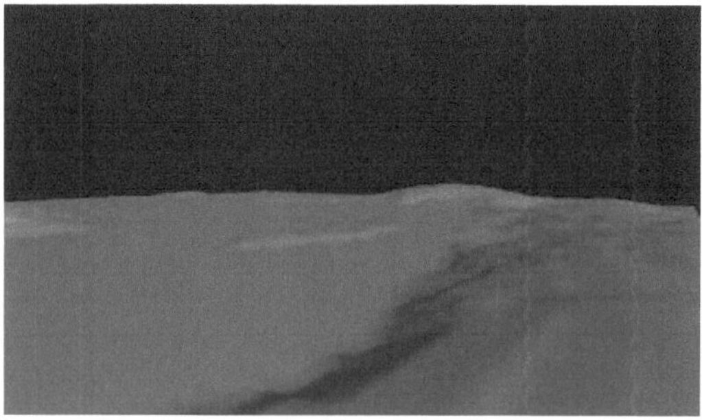

Figure 8.2 - Terrain generated using multifractal function with noise damping coefficient inversely proportional to height variation.

Another type of formation that can be created consists of terrain with reliefs eroded by erosion (such as hills). In this case, the implementation is simpler, as it is enough to reduce the number of fractal iterations, making the noise small. The relief in Figure 8.3 was generated using the standard multifractal function, with only 2 fractal iterations.

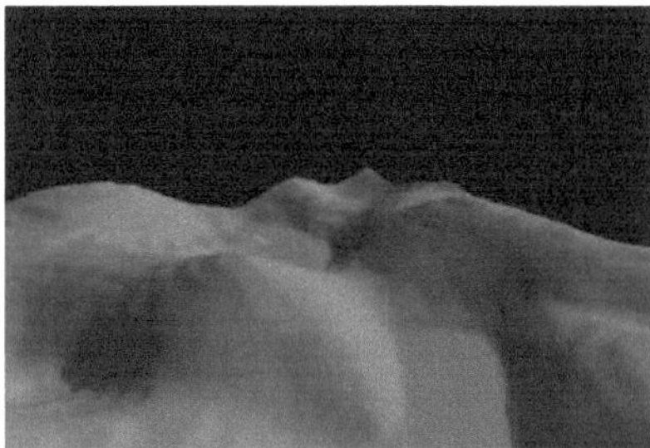

Figure 8.3 - Terrain generated using multifractal function with a small number of fractal iterations.

Numerous variations of the multifractal functions can be constructed, following the idea of impressionistic modelling. The following terrain was obtained using the function used to generate the relief of Figure 8.4 but using the idea of the turbulence function described by Ken Perlin. Here the basic function returns a value that can be negative. When this happens we convert the value to its corresponding absolute.

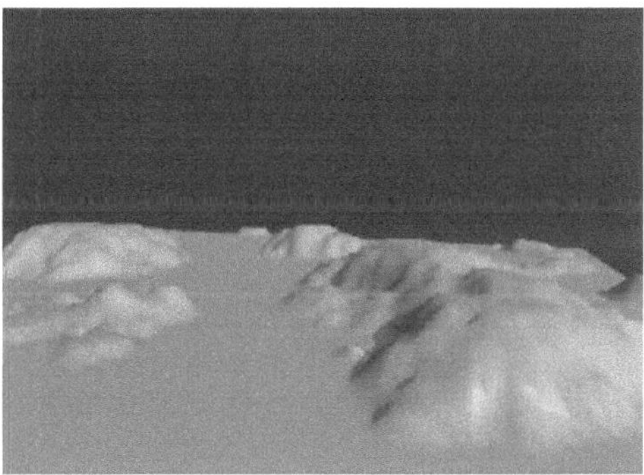

Figure 8.4 - Terrain generated using multifractal function using the idea of turbulence function, converting the negative values into their corresponding absolutes.

Many terrains are strongly determined by the rock formation they have undergone. [Musgrave 99] presents a technique for simulating terrain which is of volcanic origin, making it possible to add considerable realism to some scenes of nature. The basic idea consists of distorting the domain of the function generating the terrain using another function, which may or may not also be of fractal origin.

10.2 - WATER

The main challenge in water modelling is to simulate the surface waves (if you want to create a crystal clear and completely calm mirror lake, there will be no major problems). There are a number of works that attempt to produce a physical description of the waves, taking into account water velocity, atmospheric pressure, depth of the site, and a series of information about the physical environment, resulting on the one hand in a realistic simulation, but on the other in a relatively complex and therefore computationally expensive system of differential equations. N. Foster and D. Metaxas, in [Foster et al 99], also present computational methods based on physical models to simulate fluid motions realistically.

In parallel to these, procedural methods have been developed in order to simulate waves and their motions with more simplicity and computational efficiency. In [Ebert et al 94] Musgrave describes a wave simulation technique using the fBm function which is quite simple and offers very satisfactory results. This function is applied on a mesh and this can be done in two basic ways:

a) Use the returned value of the function as a value for perturbation of the normals, without changing the modeling, however. This technique is known as *bump-mapping* and presents good results if the deformation that is sought is relatively small in relation to the entire object. This technique has a great advantage in that the calculation can be done at rendering time, with no pre-processing required.

b) Use the returned value to create an actual deformation of the object geometry. It is necessary to resort to this method when the waves are relatively large. This requires pre-processing the mesh in question before visualization.

The method that will be described consists basically in generating a procedural texture, which is then applied on the mesh in question as *bump-mapping*. By performing a relatively small fractal iteration, patterns like the ones in the following figure are obtained:

Figure 8.5 - Texture generated with 2 and 3 fractal iterations of the *noise* function, respectively.

Musgrave suggests simply applying these textures on a mesh. In this work, however, a new and successful method has been created by seeking to create a layer of several fractal textures, each with different frequencies. In fact, the number of layers will depend on the behaviour of the water you wish to model. If, for instance, you want to create an ocean, you should use many layers, with the first layers corresponding to waves with large amplitudes and the final layers corresponding to waves that have lower amplitudes and higher frequencies. In the case of modelling a small lake, only layers with high frequencies will be used.

For this, it is first necessary to pre-calculate a table that will limit the influence value for each layer, in order to avoid a saturation in the texture that will be created. Thus, for example, assuming that in the final texture, the largest value to be mapped for each pixel is 1, and knowing that 3 layers will be used, then the first could be limited to 0.5, the second to 0.35 and the third to 0.15. Doing this, it is guaranteed that when adding the three layers, the largest possible value will be 1. To pre-calculate this table of limitations, the following algorithm can be used:

remainder = 1
  For i = 1 to (Layer_number - 1) do
        Layer_bound[i] = rest . 0,6
      remainder = remainder . 0.4
        Layer_bound [n] = rest

Finally, to create the composition of the layers, you must allocate a *buffer* with the size of the texture to be created and perform the following iteration:

    For k = 1 to Number_of_Layers do
          For each P of the buffer do
              Buffer(P)=fBm(P,n) . Layer_boundary [k] + Buffer(P)

where P is the (x,y) coordinate of the buffer points, *fBm* is the function mentioned above, returning a value between 0 and 1, and *n* is the factor for the *noise* function (high values for *n* generate a low frequency, while low values generate a high frequency).

To create an animation, it is necessary to simulate the movement of waves. In practice, a variation in amplitude over time must be performed, as well as its displacement in some direction, according to wind direction, for example:

To perform the wave displacement, you can move the point P *on* the surface, adding a direction vector, before calling the fBm function for this point. To vary the wave amplitude, you can call any function (*noise, turbulence* or even fBm itself) and multiply the original amplitude value by the result. As point P is being displaced over space, the amplitude value returned for the same point at different times will also be different. The following code summarizes this idea:

```
P = P + t . Direction_Movement
 Wave_Height = amplitude * fBm (P)
  wind_freq = 0.1
   freq_Amplitude = 1
    wind_min = 0.3
     turbulence = fBm (P * freq_wind)
       wind = wind_minimum + freq_Amplitude * turbulence
         wave = point height + wind * wave_height
```

For a more realistic simulation, this calculation should be done for each texture layer that is being used. Note that in this case, each layer should have its own values for amplitude, wind direction, wind frequency, etc.

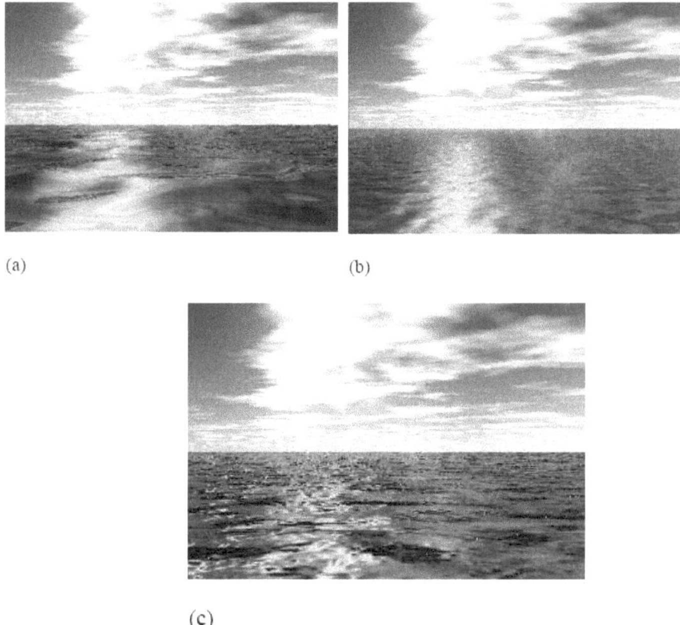

(a)

(b)

(c)

Figure 8.6 - (a) mesh using low frequency texture, with one layer generated with fBm and applied with deformation of the geometry. (b) mesh using high frequency texture, with fBm and applied as *bump-mapping* on the geometry. (c) mesh using texture with 4 layers, with frequencies varying from that used for (a) to that used for (b) and applied as *bump-mapping* on the surface.

The described method generates very realistic images of waves. However, for generating animation and some more sophisticated phenomena, such as wave breaking, this technique is not as efficient. [Tessendorf 99] presents a procedural approach for ocean simulation, based in part on real oceanographic models. For this, the author makes use of Gerstner waves [Fournier 86], which consist of equations used for approximations in fluid dynamics simulation. This model basically describes the circular motion that individual particles will possess on a surface with waves. It is defined for this particle, in the XZ plane, that $PX = (x0, y0)$ and, with respect to height, that $p_y = 0$. Thus, the displacement of particle P when a wave with amplitude A passes through it will be given by the equation below:

$$X = x_0 - \left(\frac{K_x}{k}\right).A.sen(K_x.X_0 - wt) \; e \; Y = A.cos(K_x.x_0 - wt)$$

where K *is* called a wave vector, which is horizontal with respect to the direction in which the waves move and with magnitude k, related to the wavelength ⬚⬚da from the following formula:

$$K = 2.\pi/\lambda$$

Figure 10 shows what the motion of this particle looks like with waves of various amplitudes and following the equations described previously.

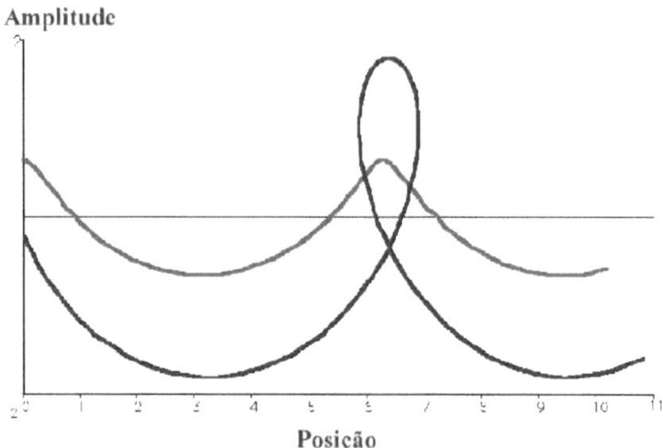

Figure 8.7 - Motion of a particle on a surface following Gerstner's model for two different values of amplitude.

Gerstner's equations, however, will only create realistic wave models when they are put into a summation of equations with various amplitudes, frequencies and *K-wave* vectors. The final summation will be as follows:

$$X = x_0 - \sum_{i-1}^{N} (\frac{K_i}{k_i}).A_i.\text{sen}(K_i.x_0 - w_i t + \phi_i) \qquad Y = \sum_{i-1}^{N} A_i.\cos(K_i.x_0 - w_i t + \phi_i)$$

Figure 9 shows the profile of a wave generated with this type of equation:

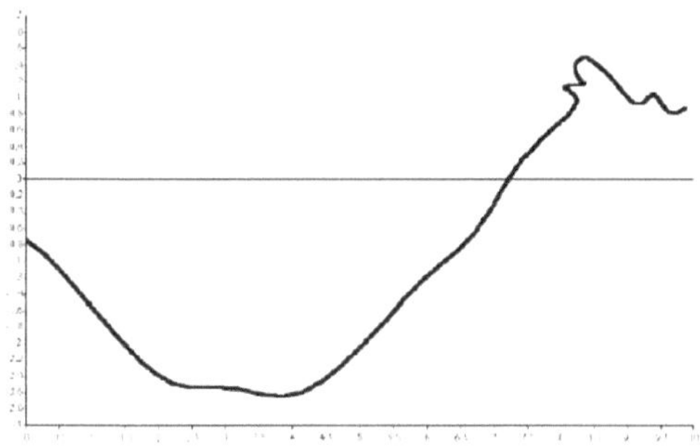

Figure 9 - Profile of a wave generated with summation of multiple equations of Gerstner

Based on this model, more realistic wave motions can be created, especially for those with a relatively large amplitude. For a more detailed study of this method, see [Tessendorf 99] and [Foster et al 99].

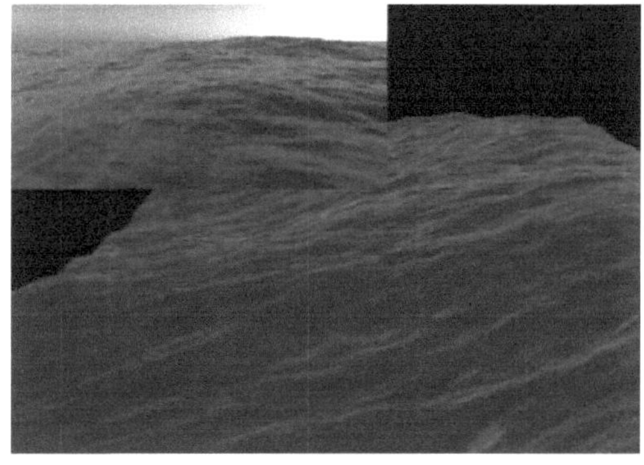

Figure 9.1 - Surface generated with summation of multiple Gerstner equations

10.3 - CLOUD MODELING

Among all the models of nature, clouds are perhaps the most complex elements to model and visualize, mainly because they do not have a well-defined surface. There are several approaches that can be made in the construction of these elements:

Gerar procedural textures, which will be used later to map them onto some object;

Gerar a three-dimensional model of a cloud layer;

Gerar the three-dimensional model of a cloud.

10.3.1 - Cloud texture synthesis

The first relevant results obtained in the attempt to synthesize cloud images procedurally are attributed to Gardner [Gardner 85] and Ken Perlin [Perlin 85]. Both observed that by directly applying their fractal functions on a two-dimensional matrix, the image formed resembled a sky with clouds. Gardner used a cosine Fourier series [Gardner 94] and Perlin used his *Turbulence* function. Later, Musgrave [Ebert et al 94] and Worley [Worley 96] also synthesised interesting images using *fBm* and fractal iteration of the basic cellular distance-based function. Once the appropriate fractal function has been constructed, synthesising a cloud image becomes quite straightforward by simply mapping the colours correctly. To generate figures 9.1 , (a) to (e) the Turbulence function was used, applied as follows:

For each point P = (x, y) of the image do
 *Color_Pixel(x, y) = (255, 255, Turbulence (x, y) * 255)*

where Color_Pixel is of the form (R, G, B)

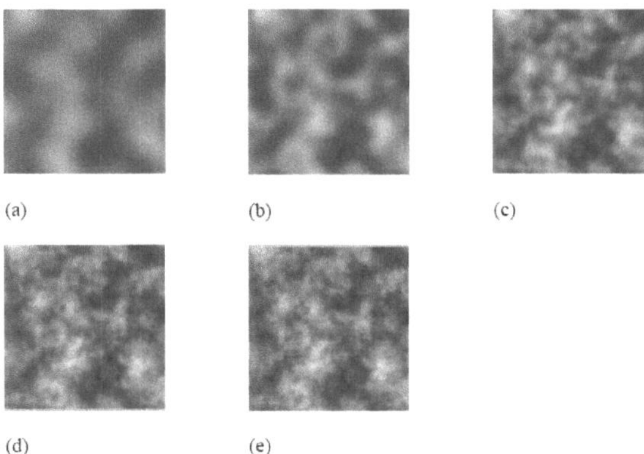

(a)   (b)   (c)

(d)   (e)

Figure 9.2 - Mapped Turbulence Function with white and blue colours, with several fractal iterations: (a) 2 fractal iterations (b) 3 fractal iterations (c) 4 fractal iterations (d) 5 fractal iterations (e) 8 fractal iterations.

10.3.2 - Three-dimensional models of cloud layers

The following is an innovative proposal for the construction of cloud layers, making use of hypertextures, presented in section 2.2.2 [Clua et al 98]. According to this technique a *fuzzy* region is initially defined, which is, in the volume of the scene, a parallelepiped aligned with the axes and defined by 2 vertices:

v1 = (0, Clouds Initial Height, 0)

v2 = (width, Final height of clouds, depth)

All clouds will be inside this rectangle, which will be divided into three horizontal slices. The lower slice is called the *lower cloudy area*, the middle slice the *central cloudy area* and the upper slice the upper cloudy *area*.

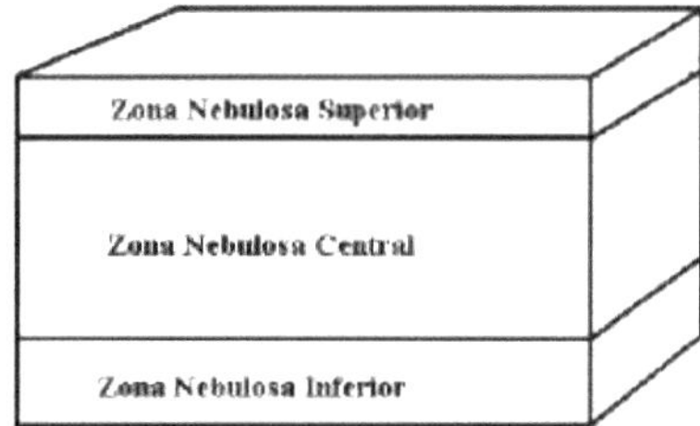

Figure 9.3 - Various *fuzzy* regions for atmospheric cloud layers.

Normally, the cloud generation process should be the last step in the creation of a scene, as it usually happens when working with hypertextures, because priorities should be given to other elements that had already been occupying a space previously, not allowing the cloud to replace them (a piece of a mountain, for example, cannot cease to exist because of the insertion of a cloud).

For each *voxel* within the central cloud region, the appropriate fractal function is used (in this work, the fBm), and the value provided as a return corresponds to the density of this point.

Note that if the extreme zones (upper and lower) are of thickness 0, the cloud layer will end abruptly, giving an unreal appearance. These zones, therefore, have a different treatment, in order to dampen the encounter of a region with clouds with another without clouds. Thus, for each *voxel* that is in the lower cloudy zone or the upper cloudy zone, the following treatment can be given to its respective density δ :

$$\delta = \left(\frac{d}{\Delta Z}\right)^{n} x fBm\,(ponto)$$

where *d is* the distance from the point to the central zone, $\varepsilon\ \varepsilon\ Z$ is the layer thickness of the zone where the point is located and *n* is the exponent that will increase or decrease the velocity of density change as one moves away from the central zone.

Still, not so satisfactory results can be obtained with this formula, because the density drop occurs homogeneously as one moves away from the central zone. To correct this problem a random value can be added to the equation. A call to the *noise* function is well suited to this situation. Thus the equation will result in:

$$\delta = \left(\frac{d}{\Delta Z}\right)^n x fBm\,(ponto)\,x\,noise\,(ponto)$$

In general, the lower part of cloud layers tends to be flatter than the upper part, due to the rising masses of warm air. To simulate this effect, it is sufficient to place a greater thickness for the upper cloud zone than for the lower cloud zone. In addition, different *n* exponents can be placed for each of the regions.

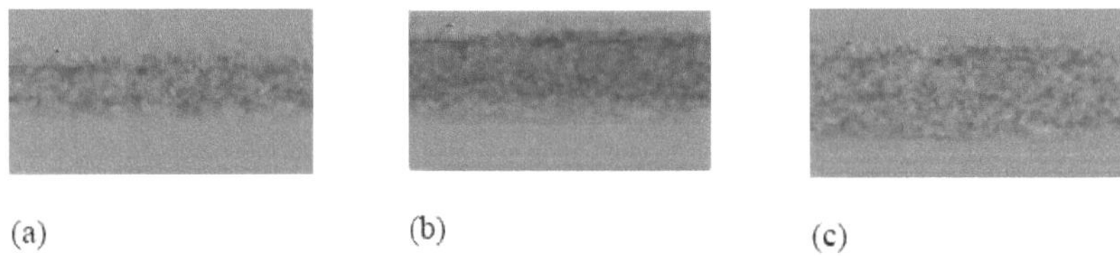

Figure 9.4 - cross-sectional slices of cloud layers placed in space. The density and thickness of each have different values.

10.3.3 - Modelling a 3D cloud

Several works have been carried out in an attempt to create realistic 3D cloud models. Neyret [Neyret 97] obtained convincing results using real physical formulations. However, this method is not very flexible and relatively complex to manipulate. Ebert, in [Ebert 97] and [Ebert 99] takes an approach using procedural functions applied to volumes, obtaining results that are also realistic and easy to manipulate for an end user. This is the method that will be addressed in this work, with some minor changes added by the author.

The modeling of a 3D volumetric cloud has two steps: the definition of its macrostructure, which will be created through implicit surfaces, and its microstructure, which will be created through procedural functions. By joining these two components, the final model of the cloud will be obtained.

Implicit functions consist of a modelling technique, that due to an attraction field created between objects by placing them relatively close to each other, models will be generated resulting from a smooth fusion between them. For a more detailed study on implicit surfaces see [Bloomenthal et al 97].

Using spheres as implicit surfaces will be a suitable method to model the macrostructure of the cloud. To model this structure, it is enough to position the centre of the spheres in space, in order to give a general shape to the element. This distribution can also be done automatically, performing a distribution of random points in space, and these points correspond to the centre of the spheres.

To generate the object composed by the implicit surfaces, one must compute the density for each *voxel* due to each of the implicit functions inserted in the space. There are several functions that serve to calculate this density, being one of the most traditional, and which perfectly suits the purposes of the work, the cubic function of Wyvill [Wyvill et al 86]:

$$F(r) = -\frac{4r^6}{9R^6} + \frac{17r^4}{9R^4} - \frac{22r^2}{9R^2} + 1$$

where *r is* the distance from a point to the centre of the implicit surface and *R* is the radius of the implicit surface. In this function, r should vary only between 0 and R. If r is greater than R, then the value of R is assigned to r, in which case the result of the function will be zero.

The final result of an implicit function system will be given by the following summation [Wyvill et al 86]:

$$\delta(V_{ijk}) = \sum_{j=1}^{IS} w_j F_j(r)$$

where δ will be the final density of the *voxel* v_{ijk} relative to the system of implicit functions, described by the functions F_j, being *r the* distance of this *voxel* to the centre of the surface Fj, w_j is the weight that F_j will have and *IS* the total number of implicit functions that compose the system.

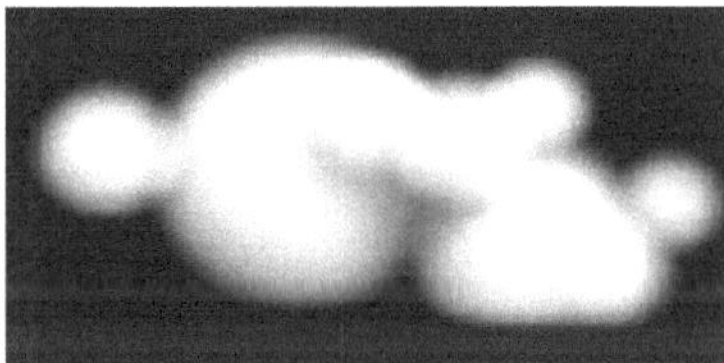

Figure 9.5 - Macrostructure created using spheres described by implicit equations.

On the other hand, the microstructure of the cloud, as seen, will be created by some procedural function (The *turbulence* function yields good results). Thus, each *voxel*

v_{ijk} *of the* interior of the volume containing the 3D cloud should have a density given by a a a a procedural function:

$$\delta (v_{ijk}) = turbulence\ (v_{ijk})$$

To perform the composition of the macrostructure with the microstructure, a combination of the result obtained in the two density calculations must be made:

$$\delta = (B \cdot \delta_1 + (1-B) \cdot \delta_2)$$

where $B \in [0,1]$ and corresponds to a clustering parameter for the two densities. To obtain more adequate results one can perturb the coordinates of the points inside the volume that will contain the cloud. For this, according to [Ebert 99], one can apply the *noise* or *turbulence* function on these points.

Volumetric_Cloud (point, Coef_Mesclage)
Point_Perturbed = Coordinates perturbed by noise or turbulence;
Density1 = Implicit Function System (Point_Perturbed);
Density2 = turbulence (Point_Perturbed);
*Density=Coef_Mescale*Density1+(1 - Coef_Mescale)*Density2*
Density = Density contrast control (e.g. exponeciation)
Return (Density); Thus, the algorithm for generating a 3D cloud can be summarised as follows:

Figure 9.6 - Cloud formed by applying the turbulence function on the macrostructure of Figure 9.5.

Through this function, a number of different cloud types can be created. This will be controlled especially by the parameter *Coef_Mesclage*, which determines whether the model is predominantly defined by the implicit surfaces or by the procedural functions and by possible changes of the procedural functions and the implicit surfaces. Thus, Ebert exemplifies in [Ebert 99] that to model a *Cumulus* cloud one should use elliptical implicit surfaces, a mixing coefficient around 40% and an exponentiation factor around 0.5 on the density of each *voxel*. For *Cirrus* clouds one should use few spherical implicit surfaces, a mixing coefficient around 30% and a small density for each *voxel*.

This type of cloud modelling allows the user to create clouds with specific shapes, such as a letter or an animal, by simply positioning the implicit surfaces appropriately.

Figure 9.7 - Simulation of a *cirrus* cloud using the parameters described above.

To generate animations of these clouds, the process is also simple. To perform global transformations (scale, rotation and translation), just apply them directly on the implicit surfaces. To perform internal animations to the model (such as simulating the "birth" of a cloud), simply make changes in the parameters of the procedural functions and density values, or even using 3D software, such as 3DS MAX, Maya, Blender, etc.

10.3.4 - Gases

Gas modelling is actually a generalisation with respect to cloud modelling. It can be used when you want to create different types of smoke, fog or even some atmospheric effects.

To create this type of elements, as with clouds, volumetric modelling, based on hypertextures (see section 2.2.2) must be used, and to carry out the visualisation, specific lighting models must be used (see section 5.1).

There are a number of techniques for giving "shape" to gases. [Ebert et al 94] describes a generic potentiation function to perform this calculation:

$$density(V) = (F(turbulence(V)) \cdot K1)^{K2}$$

This function performs a perturbation on the coordinates of *voxel V* using the *turbulence function* and then applies some convenient F function to the result obtained. $K1$ is a coefficient that will be multiplied over the result in order to be able to scale the density and $K2$ is the exponentiation constant for the expression. The model created will depend on the function F and the coefficients $K1$ and $K2$ chosen. Note that if the density value is to be restricted to the interval [0,1] of the reals, then $F(x) \in [0,1]$ and $K1 \leq 1$.

For example, to create a fog, [Ebert et al 94] suggests using the following parameters: $F(x) = x/MAX\_TURB$, $K1 = 0.5$ and $K2 = 3$, where MAX_TURB is the largest value that *turbulence* can generate in the volume being used.

To create a simple smoke column, you can use as parameters $F(x)=sen(x)$, $K1 = 0.57$ and $K2 = 6$.

From this definition, a series of changes and adjustments can be made. The smoke column mentioned above, for example, has a drawback: regardless of the height, the average density of the smoke is the same. A more realistic column should have the density decreasing as the smoke rises. More detail can also be provided regarding the objects that delimit the volumes. Instead of simple cubes, more sophisticated objects can be used.

10.3.5 - Computational fluid animation

Fluid animation occurs through a method, known as Computational Fluid Dynamics (CFD). This method is based on the *Smoothed Particle Hydrodynamics* (SPH) technique. The fluid behaviour is modelled by a system of partial differential equations (Navier-Stokes equations), subject to certain initial and boundary conditions. The SPH is applied to discretize these equations for computer simulation of the fluid motion. The result is a representation through a system of particles moving under the influence of forces arising from gravity, pressure and viscosity. Combined with efficient fluid visualisation techniques, a high degree of realism can be achieved, since the method is based on physical principles to represent fluid motion. Preliminary results taken from the literature indicate that this technique is promising.

Applications involving Computational Fluid Dynamics (CFD) have been increasingly used in several areas. In the film industry, for example, fluid modeling is the focus of research in order to increase the realism of animations. Computational Fluid Animation" can be defined as the generation of a sequence of digital images containing moving fluids. In this context, there is an interest both in faithfully reproduce scenes involving phenomena present in the real world and to produce satisfactory effects in the realization of scenes that are fruit of the imagination of the filmmakers. As an example of real scenes, we may mention: floods in open or closed environments, torrents, avalanches, explosions in urban and non-urban areas, fires and a myriad of others. Among the scenes that are the fruit of human imagination, the division of seas and rivers (Figure 2), fluids that take on human forms or other animate beings are good examples.

Fluid animation has a multidisciplinary nature, and its development depends on the interaction between computing and engineering professionals. To achieve the necessary degree of realism, it is necessary that the fluid equations are properly treated and that the data visualization techniques used are adequate to the applications. These techniques come from computer graphics and the fluid equations are derived from concepts of classical mechanics. Undoubtedly, the generation of visual effects for the entertainment industry represents a major motivator for research in the field of fluid animation. However, it is not the only consumer of the developed techniques. In the scientific and technological environment, for example,

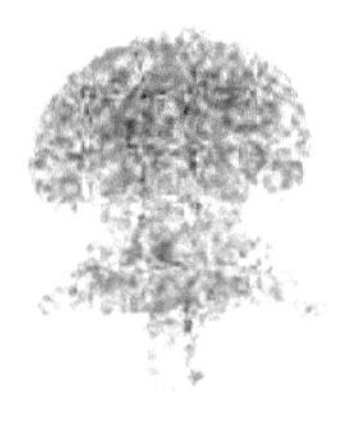

Figure 9.8: Explosion simulation using a Computational Fluid Dynamics model.

there is a growing demand for new techniques of animation, simulation and visualization of fluids, in order to apply them to the study of various phenomena. As an example, we can cite the behaviour of mechanical structures subjected to strong air currents or tidal movements, the study of atmospheric phenomena such as tornadoes, the hydrodynamics of ships, the behaviour of surfaces exposed to large temperature variations, the flow of water in certain terrains, as well as the prediction of the trajectory of volcanic lava after an eruption and the analysis of the behaviour of blood flow inside the arteries of the human body.

To study phenomena in Fluid Dynamics, researchers use physical principles together with numerical analysis resources. The fluid behavior can be modeled by a system of partial differential equations, subject to certain initial and boundary conditions. This system hardly presents an analytical solution, forcing researchers to resort to numerical analysis to discretize the problem and obtain a solution with an acceptable level of approximation. Basically, researchers have used numerical schemes based on Finite Differences, Finite Elements, as well as Lagrangean methods such as Smoothed Particle Hydrodynamics (SPH). On the other hand, distributed systems have been explored to improve the performance of applications. Such search for computational power can be attributed to the fact that, currently, a numerical simulation in Computational Fluid Dynamics can employ 3D meshes with a refinement level of the order of 256^3 points for each simulated time instant and the volume of data generated can easily reach tens of Gigabytes.

The SPH method is particle-based and has been successfully applied in engineering problems, such as compressible fluid simulation for the study of explosions and description of material dynamics. Recently, the SPH method has been explored for computational fluid animation.

10.3.6 - Works related to Fluid Dynamics

Computational Fluid Dynamics (CFD) has a long history. In 1822, Claude Navier, and in 1845, George Stokes formulated the famous Navier-Stokes equation that describes the conservation of momentum. In addition to this equation, two additional equations, one describing the conservation of mass and the other describing the conservation of energy are needed to simulate fluids. Since these equations are known and *hardware* technology exists to solve them numerically, a large number of methods have been proposed in the DFC literature to simulate fluid behaviour computationally.

For almost two decades, fluid simulation techniques for specific purposes have been developed in the field of computer graphics. In 1983, T. Reeves, introduced particle systems as a technique to model a class of *fuzzy* objects. Since then, both the Lagrangean, non mesh dependent, and the Eulerian, mesh based, approaches have been used to simulate fluids in computer graphics.

Desbrun and Cani and Tonnesen use particles to animate deformable objects. Particles have also been used to animate surfaces and to animate lava flows. In recent years, the Eulerian approach has become more popular for the simulation of general fluids, water, deformable objects and melting effects. There are few techniques available for use in interactive systems. The mesh-based method presented by Stam was certainly an important step towards real-time fluid simulation. Interactive animation techniques are also available for the special case of ocean waves. Water waves have been modelled using a wide variety of approaches, including stochastic models and kinematic models. Particle systems have been added to kinematic models to model foam and sprays to obtain breaking wave effects among others.

11 - COMPUTATIONAL FLUID DYNAMICS

Fluid dynamics is governed by the conservation laws of classical physics, called conservation of mass, conservation of energy and conservation of momentum. Partial differential equations are derived from these laws and, simplified under appropriate circumstances. Conservation of mass is assured by the "Continuity Equation" given by:

$$\frac{\partial \rho}{\partial t} + \vec{\nabla}.\,(\rho \vec{v}) = 0$$

where $\vec{v}$ is the fluid velocity.

The conservation equation for energy has the following differential form:

$$\frac{\partial}{\partial t}(\rho E) + \vec{\nabla}.\,(\rho \vec{v} E) = \vec{\nabla}.\,(k\vec{\nabla}T) + \vec{\nabla}.\,(\bar{\bar{\sigma}}.\vec{v}) + \rho \vec{f_e}.\vec{v} + qH$$

where E is the total energy per unit mass, $\square\square$é the coefficient of thermal conductivity, T is the absolute temperature and $(\bar{\bar{\sigma}}.\vec{v})$, $\vec{f_e}$ e q_H are sources of internal energy change due to internal stresses, volume forces and heat sources, respectively.

The Navier-Stokes Equation is the equation of conservation of momentum, or equation of motion, which for this paper will be written in the simplified form below:

$$\rho\left(\frac{\partial \vec{v}}{\partial t} + \vec{v}.\vec{\nabla}\vec{v}\right) = -\vec{\nabla}p + \rho g + \mu \vec{\nabla}^2 \vec{v}$$

where g is an external force field and μ the viscosity of the fluid.

The mathematical models obtained through the conservation equations described above hardly have an analytical solution. Thus, the models need to be discretized in order to perform numerical simulations. The result of this discretization is the replacement of the initial *continuum of* the model by a discrete and finite set of points in space and time, in which the derivatives that appear in the equations will be approximated.

Two approaches can be adopted to formulate the discrete versions of the model equations: the Eulerian and the Lagrangean. The Eulerian approach can be exemplified by the traditional Finite Element and Finite Difference methods, where the set of points is connected following a convenient topology, giving rise to a mesh that can be of two types:

Structured Mesh: Each internal point has the same number of neighbouring points.

2. Unstructured Mesh: Each internal point has a variable number of neighbouring points.

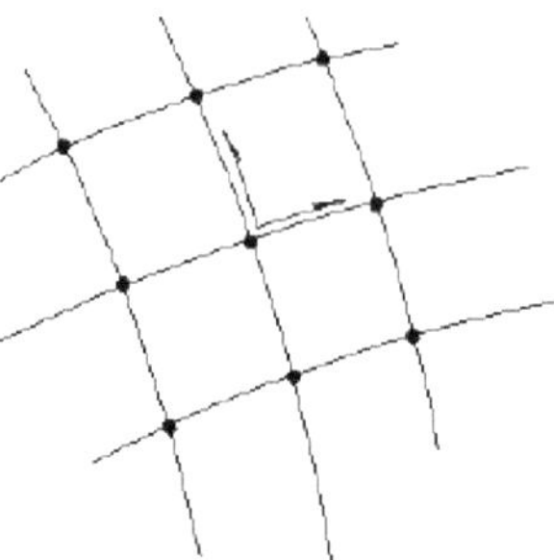

Figure 9.9 - Structured Mesh

The Lagrangean approach is mesh independent. In this case, instead of meshes, particles are used to formulate the discrete versions of the models. The method adopted in this topic for fluid animation is based on the Lagrangean approach.

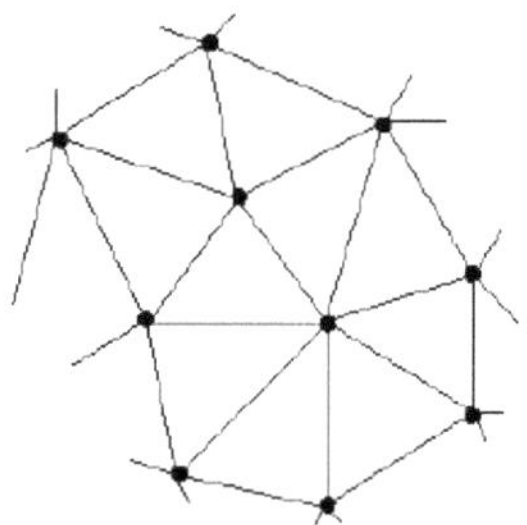

Figure 10 - Unstructured Mesh

11.1 - SMOOTHED PARTICLE HYDRODYNAMICS (SPH)

This method was developed for the simulation of astrophysical problems. Unlike mesh-based approaches, the particle-based approach has inherent conservation of mass, making it dispensable to use the continuity equation in the numerical scheme. Another interesting aspect of this approach is that particles can be directly used to "render" the fluid surface.

The fundamentals of the SPH method lie in interpolation theory. With SPH, field values that are defined only at the (discrete) positions of the particles can be evaluated at any position in space via interpolation. For this purpose, SPH distributes the quantities in a local neighbourhood to each particle through the use of a function W, called Smoothing Kernel. According to the SPH method, a scalar quantity A is interpolated at position r by a weighted sum of the contributions of all particles in its neighborhood:

$$A_s(r) = \sum_j m_j \frac{A_j}{\rho_j} W\left(r - r_j, h\right) \qquad (*)$$

Where j iterates over all particles, m_j is the mass of particle j, r_j is the position, ε_j the density and A_j the value of the magnitude at position r_j.

The function $W(r,h)$ is the smoothing kernel and should be normalised as follows:

$$\int W(r,h)\, dr = 1$$

where the constant h *is* used to define the support of the smoothing kernel.

The mass and density of the particle appear in Equation (*), since each particle i represents a given volume $V_i = m_i / \varepsilon_i$. While the mass m_i is constant for the whole simulation and, in this specific case, the same for all particles, the density ε_i varies and must be calculated at each instant of time. Particularizing Equation (*) for the density case we obtain:

$$\rho_s(r) = \sum_j m_j \frac{\rho_j}{\rho_j} W\left(r - r_j, h\right) = \sum_j m_j W\left(r - r_j, h\right)$$

In the fluid equations presented above, derivatives of the quantities involved appear. In the SPH method, it can be shown that such derivatives can be calculated from derivatives of the smoothing kernel. In particular, the gradient of A is simply:

$$\nabla A_S(r) = \sum_j m_j \frac{A_j}{\rho_j} \nabla W(r - r_j, h)$$

While the Laplacian of A is given by:

$$\nabla^2 A_S(r) = \sum_j m_j \frac{A_j}{\rho_j} \nabla^2 W(r - r_j, h)$$

It is important to realize that the SPH method presents some problems. When it is used in the fluid equations, for example, it is not guaranteed that these equations satisfy certain physical principles such as symmetry of forces and conservation of momentum. But that is another research. So far, we can stay with this method, since we will not delve into other follow-ups related to the nature of fluids. What we need to understand is their behavior in the midst of modeling from software aimed at fluid modeling in 3D environment:

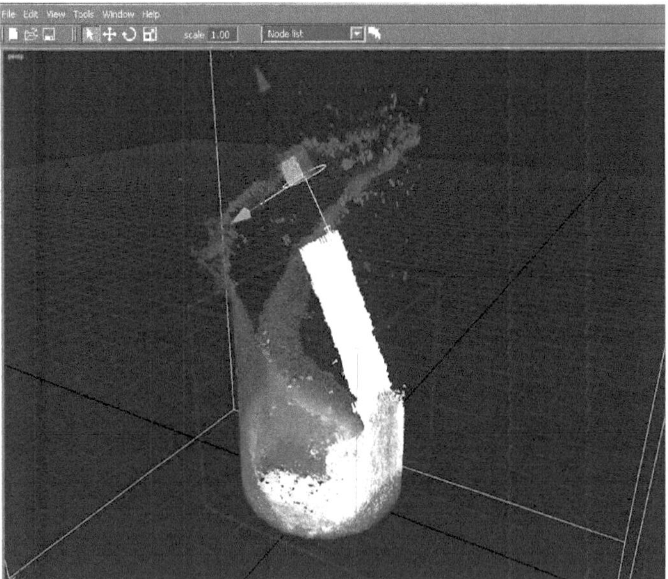

Figure 10.1 - Fluid Modeling with Real Flow 4 software

Figure 10.2 - Final render after fluid modelling

It is worth remembering that for a good result, the following script was used inside the software in Python* language, specific to some 3D software:

```
1.   # Name :pressure_multi_spray
2.   # Description: Creates Dumb Particle spray on collision based on
 pressure of the colliding particle in multiple spray emitters.
3.   #
4.   # Author      : R. Damasclin

5.   def onSimulationStep():
6.   pass

7.   def onSimulationFrame():
8.   from random import random
9.   #Emitters, add/uncomment as needed.
10.  water = scene.getEmitter("water") #Name of the main fluid emitter.
11.  spray = scene.getEmitter("spray") #name of the spray emitter
12.  #sprayb = scene.getEmitter("sprayb") #name of the spray emitter
13.  #sprayc = scene.getEmitter("sprayc") #name of the spray emitter
14.  #sprayd = scene.getEmitter("sprayd") #name of the spray emitter
15.  #spraye = scene.getEmitter("spraye") #name of the spray emitter
16.  #sprayf = scene.getEmitter("sprayf") #name of the spray emitter
17.  #sprayg = scene.getEmitter("sprayg") #name of the spray emitter
18.  #sprayh = scene.getEmitter("sprayh") #name of the spray emitter
19.  #Control Variables
```

```
20.  spray_pressure=10000 #The maximum pressure of the colliding
  particle to create spray (More = Less Spray)
21.  spray_parts = 2 #number of spray particles per collision/emitter
22.  coll_skip=4 #how many collisions to skip
23.  spray_rand=15 #how random the direction is, this depends on the
  speed of the particles.  If you expect them to be
24.  #going very fast, or want a wider cone of spray then increase this
25.  spray_speed_rand=3 #how much extra speed is added to the particles
  in the existing velocity vector
26.  #housekeeping
27.  h_spray_rand=spray_rand/2
28.  part_count=0
29.  #loop through Particles find collision's and emit spray.
30.  particles = water.getParticlesColliding()
31.  for particle in particles:
32.  part_count=part_count+1
33.  if(part_count % coll_skip == 0):
34.  press=particle.getPressure()
35.  if(press < spray_pressure):
36.  pos = particle.getPosition()
37.  vel = particle.getVelocity()
38.  tx=vel.getX()
39.  ty=vel.getY()
40.  tz=vel.getZ()
41.  for i in range(spray_parts):
42.  vel.setX(tx * (1 + (random() * spray_speed_rand)) + (random() *
  spray_rand) -h_spray_rand)
43.  vel.setY(ty * (1 + (random() * spray_speed_rand)) + (random() *
  spray_rand) -h_spray_rand)
44.  vel.setZ(tz * (1 + (random() * spray_speed_rand)) + (random() *
  spray_rand) -h_spray_rand)
45.  spray.addParticle(pos, vel)

46.  ##uncomment / copy one of these for EACH spay emitter.
47.  #for i in range(spray_parts):
48.  #vel.setX(tx * (1 + (random() * spray_speed_rand)) + (random() *
  spray_rand) -h_spray_rand)
49.  #vel.setY(ty * (1 + (random() * spray_speed_rand)) + (random() *
  spray_rand) -h_spray_rand)
50.  #vel.setZ(tz * (1 + (random() * spray_speed_rand)) + (random() *
  spray_rand) -h_spray_rand)
51.  ##SET THE NAME BELOW TO sprayc, d, e, etc.
52.  #sprayb.addParticle(pos, vel)
53.  def onSimulationBegin():
54.  pass
55.  def onSimulationEnd():
56.  pass
57.  def onChangeToFrame():
58.  pass
```

***Python** is a high-level programming language interactive, interpreted object-oriented and typing language dynamic and strong typing, launched by Guido van Rossum in 1991.[1] It currently has a community and open development model managed by the non-profit organization Python Software Foundation. Although various parts of the language have formal standards and specifications, the language as a whole is not formally specified. The standard *in fact* is the CPython.

The language was designed with the philosophy of emphasizing the importance of programmer effort over computational effort. It prioritizes code readability over speed or expressiveness. It combines a syntax syntax with the powerful features of its library library and by modules and *frameworks* developed by third parties.

The name **Python** originated from the British comedy group group Monty Python, creator of the show *Monty Python's Flying Circus*although many people make association with the reptile of the same name (in PortuguesePython). - Source: **http://pt.wikipedia.org/wiki/Python**

11.2 - FLUID MODELLING RESULTS

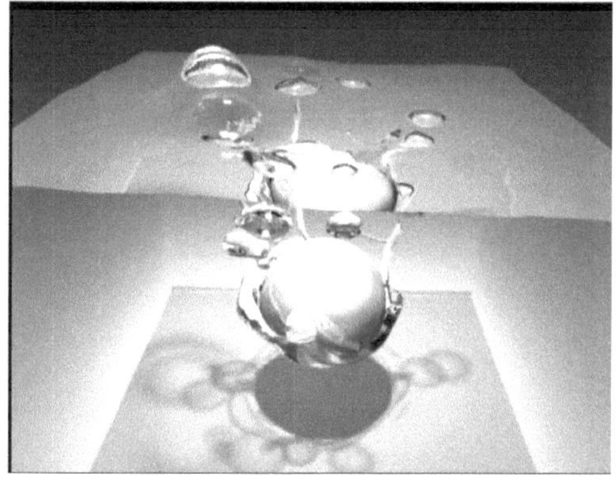

Figure 10.3: Creating a sphere within a "liquid" solution producing fluid dynamics

Figure 10.4: Render after modelling a fluid and its dynamics

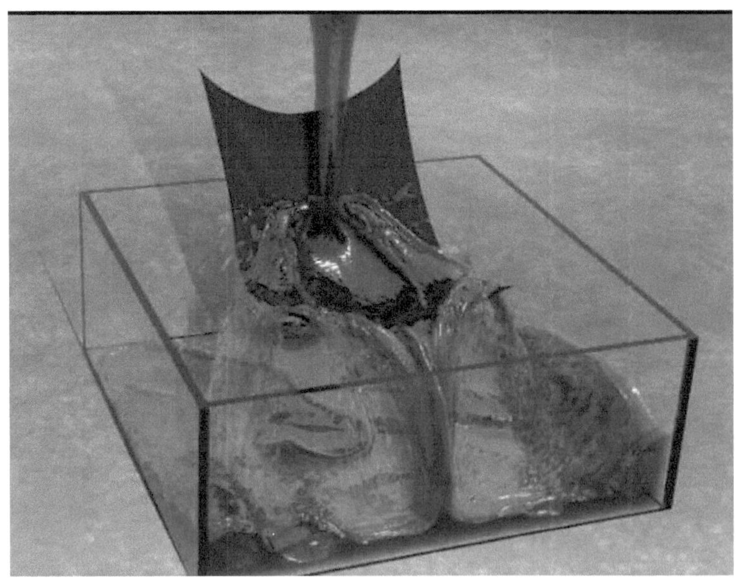

Figure 10.5 - Fluid and friction: Dynamics between surface

10.6

10.7

10.8

10.9

In the figures above, we can verify the initial instant (10.6) of a cubic object, where it "suffers" alteration in its structure (10.7 / 10.8 / 10.9), through some algorithms, an "explosion" was produced. This fractal that produced this gas describes a generic potentiation function. Where this function performs a perturbation in the coordinates of the *voxel* producing a *turbulence*.

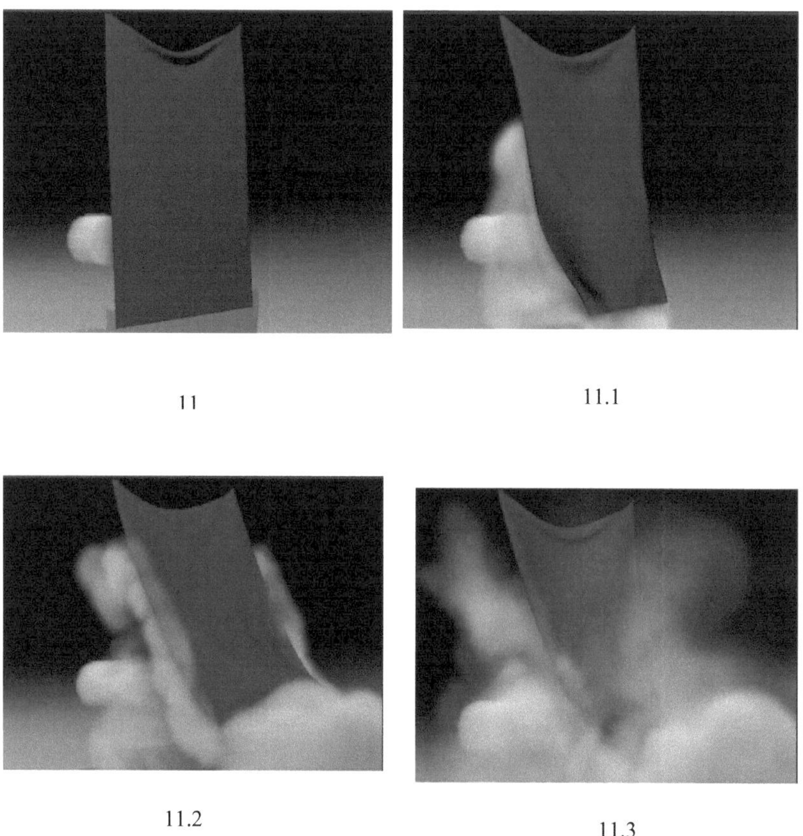

11   11.1

11.2   11.3

In the figures above, we can also verify, the initial instant (11) at which the gas, generated from a fractal, "pushes" a "cloth", where we can verify the dynamics of this gas. As in figures 11.1, 11.2 and 11.3. The same algorithm, suffering some changes, is able to describe a perturbation in the coordinates of the *voxel* producing a *turbulent* dynamics.

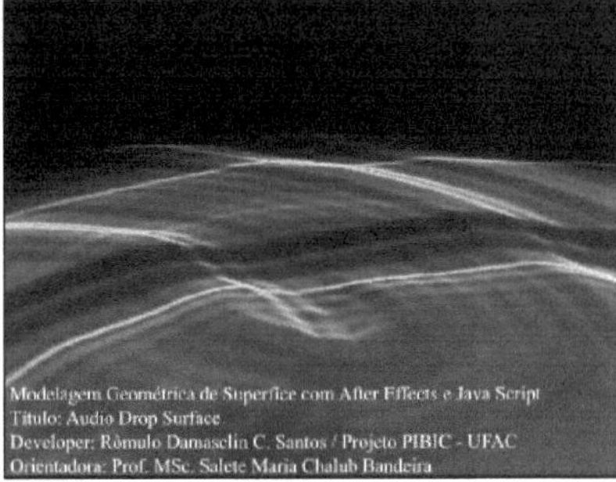

11.4   11.5

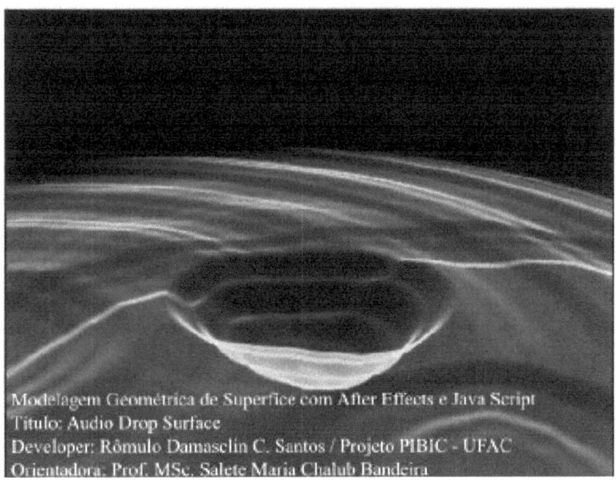

11.6

In the figures above, we can also verify the initial instant (11.4) at which the fluid, generated from a fractal, "propagates" over a surface, where we can verify its dynamics. The same algorithm, with a few changes, is able to describe a perturbation in the coordinates producing a *turbulence* function. This framework was fully developed using Adobe After Effects software along with JavaScript in the wave propagation orientation.

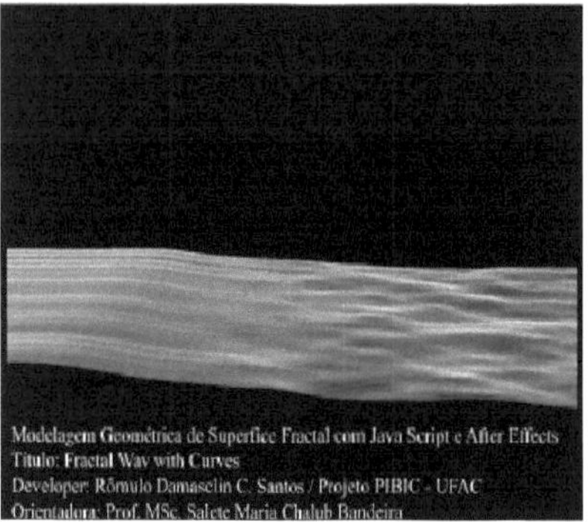

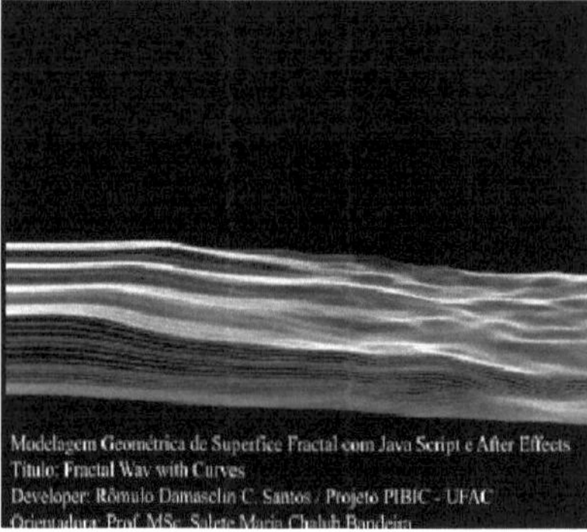

11.7   11.8

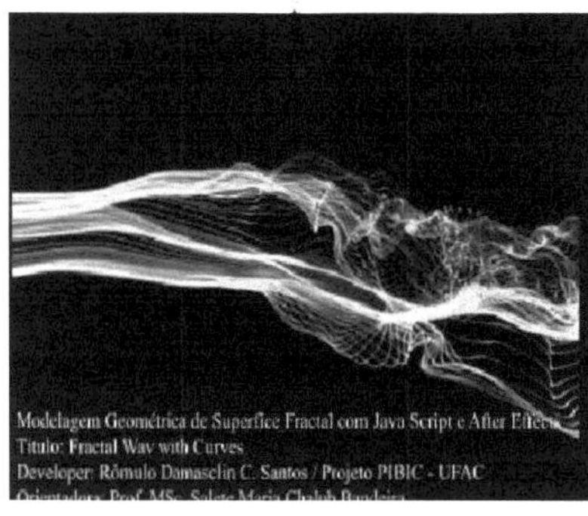

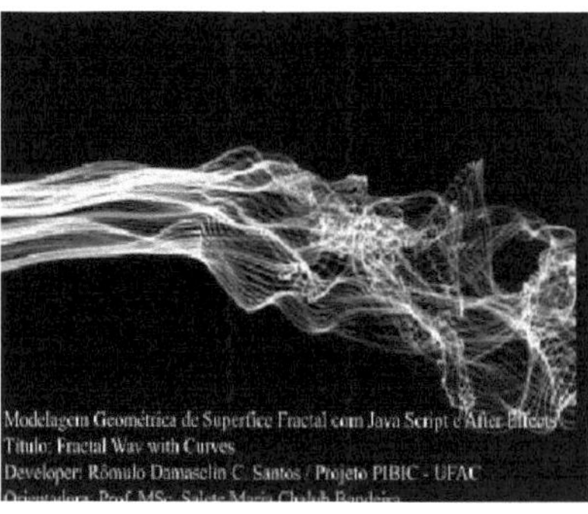

11.9   12

In the figures above, we can also verify the initial instant (11.7) at which the surface, generated from a fractal, undergoes a perturbation in its structure. With this, it produces dynamics and changes in its structure. This structure was fully developed using the Adobe After Effects software along with JavaScript in the deformation of the surface meshes.

CONCLUSION

The study of Geometric Modeling of Fractals with Object Oriented Language reveals itself so far, only as the "tip of an iceberg". Characterized of a great importance for the knowledge areas. The geometric modeling of fractals has relevant methods, when it comes to modeling "objects" of nature, such as fluids, terrain, clouds, etc.. The geometry of chaos, although a subject not much explored in our environment, but only addressed in a superficial way, becomes to have an importance of great interest in solving problems related to the study of climate and weather, the study of the behavior of fluids and their dynamics, among other problems where we can not find exact and precise solutions. After this study we start to look at the non-Euclidean geometry as a geometry that brings in its structure, answers to various problems not yet solved by a more accurate and rectilinear geometry.

Some applications are quite relevant in the study of fractals, for example, in Geography the description and characterization of seismic faults and, consequently, earthquakes are obtained through the study of their fractal structure. Besides earthquakes, other geological phenomena can be studied such as, for example, the dynamics of volcanoes. In economics, the analysis of global behaviour helps to understand local behaviour and vice versa. The study of these analyses makes it possible to create medium and long-term commercial strategies.

Another quite important result is the multidisciplinary nature of fluid animation. It is another aspect to be highlighted. The development of this area depends on the interaction between the areas of computation and fluid mechanics. This fact brings a singular dimension to this area of computational animation. On one hand, animation specialists seek in the fluid modeling techniques, tools to meet their needs in computer graphics. The final rendering of the scenes depends on the development of resources in scientific visualization which, in turn, may be useful to engineers in the analysis of the results of their simulations. Therefore we can glimpse a very fertile field of work with ample possibilities for research.

BIBLIOGRAPHIC REFERENCES

BOLYAI, John, *The Science of Absolute Space*, English translation by George Bruce Halsted, in BONOLA, Roberto, *Non-Euclidean Geometry*, English translation by H. S. Carslaw; New York: Dover Publications Inc, 1955.

BONOLA, Roberto, *Non-Euclidean Geometry,* English translation by H. S. Carslaw; New York: Dover Publications Inc, 1955.

DAVID, Ebert. ***Texture and Modeling***: ***A procedural Approach***. Academic Press, 1994.

______. Procedural Volumetric Modeling and Texturing, ***SIGGRAPH 97***, course 14, notes, chapter 6, August 1997.

DEVANEY, R. L., Differential Equations, Dynamical Systems, and an Introduction to Chaos, second edition, Elsevier Academic Press Inc.

FOSTER, Nick; METAXAS, Dimitri. Realistic Animation of Liquids. ***SIGGRAPH 99***, *cours notes*. August 1999.

GOODMAN, Danny. Java ***Script: The Bible. Translation: Daniel Vieira***. 3ª triagem. Rio de Janeiro: Elsevier, 2004.

GUIDORIZZZI, Hamilton Luiz. ***Um Curso de Cálculo***. 4. ed. V. 1, 2, 3 e 4. Rio de Janeiro: LTC, 1997.

LOBACHEVSKY, Nicholas, *The Theory of Parallels*, English translation by George Bruc Halsted, in BONOLA, Roberto, *Non-Euclidean Geometry*, English translation by H. S Carslaw; New York: Dover Publications Inc, 1955.

Mandelbrot, B. The Fractal Geometry of Nature.

W.H. Freeman, New York, 1983.

MUSGRAVE, F. Kenton; KOLB, Craig E. The synthesis and rendering of eroded fractal terrains. *Computer Graphics (SIGGRAPH '89Proceedings)*, v. 23, p. 41-50, July 1989.

NEYRET, F., Qualitative Simulation of Conventional Cloud Formation and Evolution. *Eight International Workshop on Computer Animation and Simulation, Eurographics*, September 1997.

Blender 3D free software. Available at: < www.bender.org > Accessed on: 20 nov. 2007.

TRICOT, C. Curves and Fractal Dimension. Springer-Verlag, New York, 1995.

WYVILL, G. ; MCPHEETERS, C.; WYVILL, B. . **Data Strucuture for Soft Objects, *The Visual Computer***, v. 2, p 227-234, January 1986.

Printed by Books on Demand GmbH, Norderstedt / Germany